Measuring Time

Frequency measurements and related developments in physics

Measuring Time

Frequency measurements and related developments in physics

Masatoshi Kajita
National Institute of Information and Communications Technology, Tokyo, Japan

IOP Publishing, Bristol, UK

ISBN 978-0-7503-2124-2 (ebook)
ISBN 978-0-7503-2122-8 (print)
ISBN 978-0-7503-2123-5 (mobi)

DOI 10.1088/978-0-7503-2124-2

Version: 20180901

IOP Expanding Physics
ISSN 2053-2563 (online)
ISSN 2054-7315 (print)

British Library Cataloguing-in-Publication Data: A catalogue record for this book is available from the British Library.

Published by IOP Publishing, wholly owned by The Institute of Physics, London

IOP Publishing, Temple Circus, Temple Way, Bristol, BS1 6HG, UK

US Office: IOP Publishing, Inc., 190 North Independence Mall West, Suite 601, Philadelphia, PA 19106, USA

Contents

Acknowledgment		**vii**
Author biography		**viii**
1	**Standards of physical quantities**	**1-1**
1.1	Introduction	1-1
1.2	The time and frequency standard	1-2
1.3	The length standard	1-4
1.4	The mass standard	1-5
1.5	From macro to micro	1-6
	References	1-7
2	**Historical role of precision measurements of time and frequency for the development of physics**	**2-1**
2.1	Introduction	2-1
2.2	Change from Ptolemaic theory to Copernican theory	2-1
2.3	Measurement of the speed of light	2-4
2.4	Theory of relativity	2-8
	References	2-12
3	**The fundamentals of an atomic clock**	**3-1**
3.1	Introduction	3-1
3.2	Attainable accuracy of atomic clocks	3-3
	3.2.1 Statistical uncertainty	3-3
	3.2.2 Systematic uncertainty	3-4
3.3	Accuracy and stability	3-7
3.4	Atomic clock operating in the microwave region	3-8
3.5	Optical atomic clocks	3-13
3.6	Precision measurement of molecular transition frequencies	3-20
3.7	Other possible clocks	3-22
3.8	Comparison between atomic clocks	3-23
	References	3-26

4	**The role of precision measurements in future developments in physics**	**4-1**
4.1	Introduction	4-1
4.2	Confirmation of theory of relativity	4-1
	4.2.1 Second-order Doppler shift and gravity redshift	4-1
	4.2.2 Detection of the gravitational wave	4-2
	4.2.3 Search for the gravitational effects at the micro-scale	4-5
4.3	Search for symmetry violations in particles and anti-particles	4-7
	4.3.1 C-, P-, and T-symmetry	4-7
	4.3.2 Confirmation of CPT-symmetry	4-10
	4.3.3 Search for electron electric dipole moment	4-11
4.4	Search for the variation in the fundamental constants	4-14
	4.4.1 Variation in the fine structure constant	4-15
	4.4.2 Variation in the proton-to-electron mass ratio	4-16
4.5	Search for symmetry violation in chiral molecules	4-18
4.6	Role of atomic clocks for astronomy	4-19
	4.6.1 History of the Universe	4-20
	4.6.2 Dark energy	4-21
	4.6.3 Dark matter	4-22
4.7	Quantum electrodynamics and proton size	4-23
	References	4-25
5	**Role of atomic clocks for human life**	**5-1**
5.1	Position measurement by GPS	5-1
5.2	Very long baseline interferometry	5-2
5.3	Establishment of a national standard time and its distribution to users	5-3
5.4	Contribution to communication technology	5-4
5.5	Monitoring air composition	5-5
5.6	Temperature measurement	5-6
	References	5-8
6	**Conclusion**	**6-1**
Appendix A: Laser cooling		**A-1**

Acknowledgment

The research activity of the author is supported by a Grant-in-Aid for Scientific Research (B) (Grant No. JP 17H02881), a Grant-in-Aid for Scientific Research (C) (Grant Nos. JP 17K06483 and 16K05500), and a Grant-in-Aid for Exploratory Research (Grant No. JP15K13545) from the Japan Society for the Promotion of Science (JSPS). The author is highly appreciative of discussions with Y Yano, T Ido, N Ohtsubo, Y Li, H Hachisu, S Nagano, and M Hara, all from the NICT, Japan, as well as T Aoki (University of Tokyo), M Yasuda (AIST, Japan), and A Onae (AIST, Japan; passed away 12 June 2017). The author is grateful to K Kameta and J Navas (IOP Publishing, UK) for the opportunity to write this book.

Author biography

Masatoshi Kajita

Born and raised in Nagoya, Japan, Dr Kajita graduated from the Department of Applied Physics at the University of Tokyo in 1981 and obtained his PhD from the Department of Physics at the University of Tokyo in 1986. After working at the Institute for Molecular Science, he joined the Communications Research Laboratory (CRL) in 1989. In 2004, the CRL was renamed the National Institute of Information and Communications Technology (NICT). In 2009, he was a guest professor at the University of Provence, Marseille, France.

IOP Publishing

Measuring Time

Frequency measurements and related developments in physics

Masatoshi Kajita

Chapter 1

Standards of physical quantities

1.1 Introduction

Physics concerns the laws of nature, from which we can make predictions and retrodictions of the behavior of phenomena. Researchers seek laws in the phenomena that they experience around them. Experimentalists confirm the laws and look for new phenomena whereas theoreticians seek unknown phenomena based on the laws or by extending them. With the discovery of new phenomena that cannot be described by established laws, new models are developed to explain them and thus we take new steps in advancing physics knowledge.

Measurements of physical quantities are important in describing phenomena in detail. For example, the empirical law 'the distance traveled by a falling body is directly proportional to the square of the time taken in the fall' was established first by Galilei through measurements of fall times and distance traveled. However, any measurement (for example, X) comes with an inherent uncertainty (ΔX), and the measurement result should be given as $X \pm \Delta X$. We must compare this measurement result with estimates given by the different models. If one model gives the value $X + \delta_1 X$ and another model gives $X + \delta_2 X$, we cannot say which model is right when $|\delta_{1,2} X| < \Delta X$. The model that is easiest to imagine for most scholars has been, on the whole, the one most accepted. Only after the uncertainty in the measurement has been reduced and $\Delta X < |\delta_1 X - \delta_2 X|$ holds, can we conclude which models to eliminate.

To discuss the uncertainty in a measurement, we must determine the standards associated with each physical quantity. Note that physics is based on three fundamental physical quantities: mass (M), length (L), and time (T). All other physical quantities are combinations of these three fundamental quantities and are given in the form $M^{n(M)} L^{n(L)} T^{n(T)}$, where $n(M, L, T)$ are the dimensions of each fundamental quantity. Several typical examples are given below:

doi:10.1088/978-0-7503-2124-2ch1

Velocity	$M^0L^1T^{-1}$	Acceleration	$M^0L^1T^{-2}$
Force	$M^1L^1T^{-2}$	Energy	$M^1L^2T^{-2}$.

Therefore, the measurement accuracy is given by the measurement accuracies to which the three fundamental physical quantities are known. The current convention is to present values of these three physical quantities using the MKS-system of units:

Mass	kilogram	(kg)
Length	meter	(m)
Time	second	(s).

There are questions as to how accurately these three fundamental quantities are determined. We need to consider the standard for each physical quantity so that the measurement of time/length/mass by any clock/ruler/solid metering device should be equal. With one clock, we read the time trusting this clock is correct. However, with two clocks displaying different times (figure 1.1), at least one clock has problems with accuracy due to mechanical problems or the initial settings. We do not know which clock, if either, is correct, but from their time differences, we can imagine the clocks have an associated uncertainty.

Next, we introduce standards by which the units of the three fundamental quantities are defined.

1.2 The time and frequency standard

The standard for time and frequency has changed over the centuries (figure 1.2). In the beginning, time was established from the solar day, one rotation of the Earth

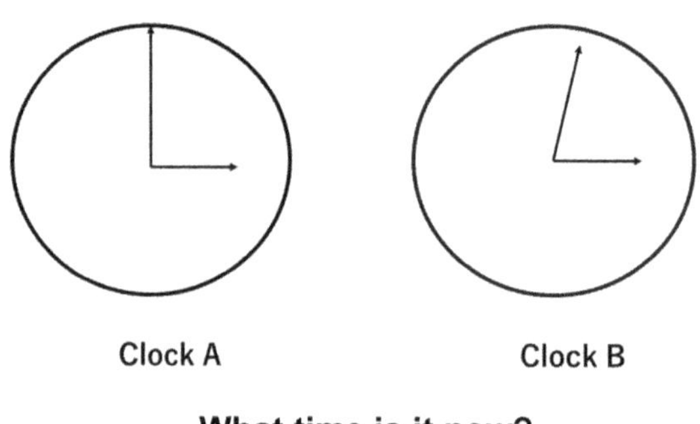

Figure 1.1. A situation where two clocks display different times. The question is which clock, if either, tells the actual time.

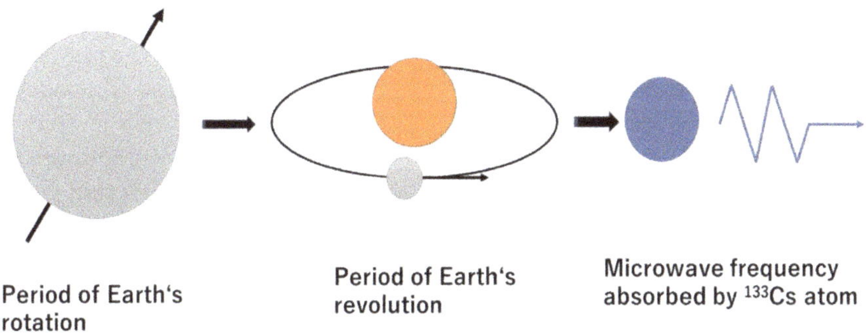

Figure 1.2. History of the standard of time and frequency.

about its axis, and was defined as containing 86 400 s. However, this standard is not always appropriate to use in practice or to describe physical laws. The reason is that the solar day is not constant because of gravitational tides, principally stemming from the gravitational interaction with the Moon. Over a time scale of a million years, Earth's rotation is slowing down. Three billion years from now, 1 day is expected to be 31 h. However, it can also fluctuate over time scales of 10–50 years by a factor of 10^{-8} [1].

During the period 1956–67, the standard of time was changed to the period of Earth's orbit around the Sun. Taking just the gravitational interaction between the Sun and Earth, and assuming their masses are constant, the orbital period should be constant. However, gravitational perturbations from other planets cause fluctuations in the orbital period. Note also that the Sun is losing mass and hence the radius of Earth's orbit about the Sun is increasing. Therefore, the orbital period is getting longer. Its fluctuation is of order 10^{-9}, which is just one order smaller than the rotation period of the Earth [2].

After World War II, atomic clocks were invented. An atomic clock is based on the frequency of the electromagnetic waves (e.g. light, microwaves, x-rays) absorbed by atoms. The attainable accuracy was improved dramatically with the introduction of such clocks. The basic principles of operation are the following.

(1) Each electromagnetic wave has a characteristic frequency f and an associated particle (called the photon) having energy hf (h is the Planck constant).

(2) Each particle has a characteristic wave associated with it. The electrons in atoms can only have orbits, called Bohr orbits, whose length ($2\pi \times$ radius) is a multiple of the electron's matter wavelength (called the de Broglie wavelength). Therefore, the energy of the electron can take only discrete values.

(3) The electrons in atoms can change their state of energy (make a transition) by the difference ΔE by absorbing or emitting a photon; for the total energy to be conserved $\Delta E = hf$. That is, the electromagnetic wave is absorbed or emitted only if the frequency satisfies this condition (called the transition frequency).

A more detailed exposition is given in chapter 3. The first atomic clocks were developed using transition frequencies that fell within the microwave region because such frequencies could be measured using counters. In 1967, the frequency of the microwave absorbed by a cesium (^{133}Cs) atom was defined to be 9 192 631 770 Hz and has become the current standard for time and frequency [3]. Its uncertainty at that time was of order 10^{-12} (the frequency difference between different Cs clocks). This clock takes 30 000 years to produce a 1 s error. Currently, the uncertainty associated with Cs atomic clocks is reduced to around 10^{-16} (i.e. it takes more than 100 million years to produce an error of 1 s) [4].

Since the beginning of the twenty-first century, a new apparatus called the frequency comb has been developed, which made it possible to measure optical frequencies. The precise measurements of atomic transition frequencies in the optical region became one of the hottest topics for many researchers. An uncertainty of order 10^{-18} was obtained with several atomic transitions [5–8]. It seemed opportune to establish a new standard of time and frequency based on an atomic transition frequency in the optical region, but deciding which atomic transition frequency is best has required some thought. A re-definition of time and frequency is to be given sometime after 2026.

1.3 The length standard [9]

In 1793, the French Academy of Science advised adopting a definition for the unit of length, called the meter, to be one ten-millions of the distance between the North Pole and the equator. In 1799, a standard bar of platinum was made based on the meridional meter length. This standard bar became known as the 'Mètre de Archives'. In 1869, the definition of the meter was changed to this 'Mètre de Archives' because it was difficult to precisely measure the size of the Earth. The uncertainty was of order 10^{-5}. In 1889, the definition was changed to the distance between two lines on a standard bar measured at the melting point of ice. The standard bar was made of an alloy of platinum with 10% iridium (the alloy being harder than pure platinum) with an X cross-section to minimize bending. The uncertainty is of order 10^{-7}.

In the 1940s, the idea to use the wavelength of light as the length standard was proposed. In 1960, a new definition of the meter length was given in terms of the wavelength of light emitted by krypton-86 (^{86}Kr); the uncertainty was reduced to 10^{-9} [10].

Earlier, in 1881, Michelson and Morley found that the speed of light c is constant for all observers regardless of speed [11]. Initially, its measurement was given with an uncertainty of 10^{-8} but has subsequently been reduced to 10^{-15} by Brillet and Hall [12]. Because of its universal constancy, a value of $c = 299\,792\,458$ m s^{-1} was set in 1983. With this exact value, the wavelength of light λ is then given by c/f. Hence, a wavelength of laser light known to have a low frequency uncertainty offers a means to provide an accurate length scale.

1.4 The mass standard

The original kilogram (kg), the base unit of mass, was defined as the mass of distilled water in a cubic container of size 10 cm × 10 cm × 10 cm (i.e. one liter of water) at the temperature (4 °C) at which water is at its densest. Its uncertainty was given by the uncertainties of the container size, the purity of the water, and the temperature. Since 1889, an artifact has defined the kilogram standard (International Prototype of the Kilogram or IPK) stored at the International Bureau of Weights and Measures (BIPM) on the outskirts of Paris [13]. It is a right-circular cylinder (diameter = height, figure 1.3) of 39 mm made from an alloy of platinum (90%) and iridium (10%). The hardness of pure platinum is enhanced by the addition of iridium; moreover, its surface area is minimized by using the right-circular shape. In addition, six copies of the IPK were made, but there is a mass difference to that of the IPK of 10–100 μg [14] because of limits in the accuracy of size and mixture rates of iridium during fabrication. In addition, mass differences have drifted over time. The masses have increased with dust in the air settling on surfaces and have also decreased from the original values after cleaning. The mass uncertainty is of order 5×10^{-8}.

The mass of an atom is always constant if relativistic effects are negligibly small. The ratios of the atomic mass to the electron mass have been measured to high accuracy; for example, the proton-to-electron mass ratio has been determined to an uncertainty of 9.5×10^{-11} [15]. In November 2018, a new proposal is to be presented to define Planck's constant and unit electric charge as exact values. If the electron mass is well known, then the atomic masses are also well-determined with high accuracy. The mass standard can also be defined with higher accuracy if we can estimate the number of atoms. A mass of 1 kg can be obtained from a 93.6 mm diameter sphere of silicon (Si) atoms (figure 1.4). Si was chosen because ultra-pure defect-free monocrystalline silicon already exists in fabricated form. The lattice spacing between atoms in the crystal, 0.192 nm (1.92×10^{-7} mm), can be measured using a scanning x-ray interferometer. The size of this Si sphere has been measured using optical interferometry to an uncertainty of 0.3 nm (3×10^{-7} mm)—roughly the

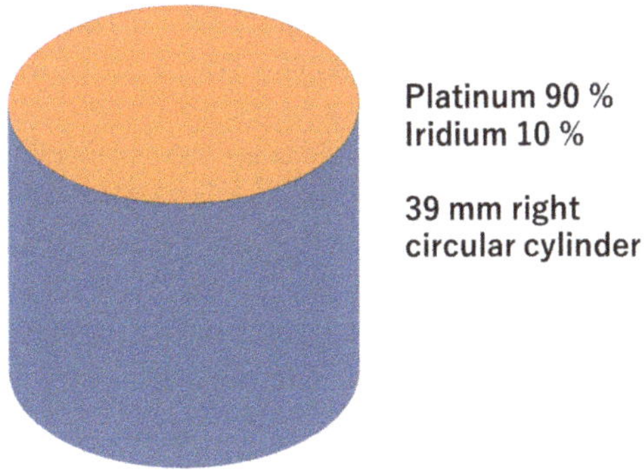

Figure 1.3. Schematic of the current mass standard.

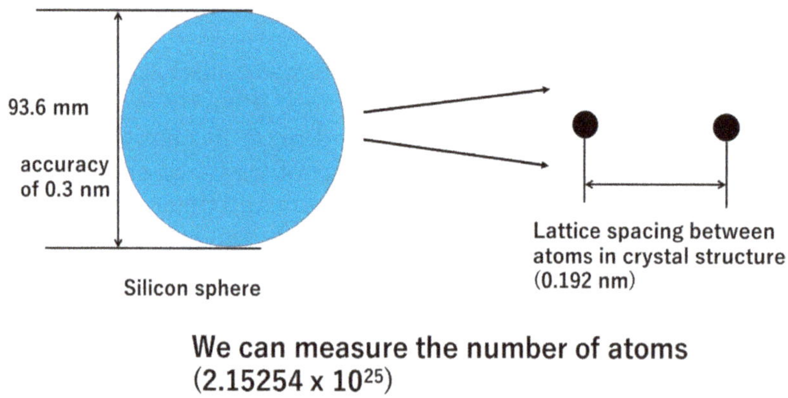

Figure 1.4. Schematic of the future mass standard, a pure Si sphere.

size of a single atom [16]. With this method, the mass uncertainty of the Si sphere is expected to be reduced.

1.5 From macro to micro

The standards for the three fundamental physical quantities were originally defined using macro-size artifacts (solid substances) and have recently been changed to micro-size objects (gaseous atoms). Clearly, if a standard is defined in terms of solid substances, we cannot make an exact prototype of the standard because there are limits to the accuracy of its size and the purity of substances.

In contrast, the structure of each species of atom is generally stable and universal. There is no difference in the measurement using different individual atoms of the species. However, there remains the question of whether the measurement of the attributes of a micro-size object can be connected to the macro-sized version of that object.

The frequency of the atomic transition in the microwave region is 10^6–10^{10} Hz, which can be measured in a comparison with frequencies of a crystal oscillator with macro size. Therefore, the definition of time and frequency given at the micro-scale size can also be applied at the macro-scale size. With the development of the frequency comb, optical frequencies can also be measured with an apparatus of macro-scale size. The overlap of the frequency region at both scales makes it possible to give a common standard used in both scales.

If the mass standard were defined with a micro-sized object, it would be much simpler to determine the mass of elementary particles. Unfortunately, the mass regions in the micro- and macro-scale sizes are different by more than 20 orders, making it difficult to have a standard applicable at the macro-scale size. Recent research has made it possible to estimate the number of silicon atoms in a sphere, and the improvement of the accuracy of the mass standard is expected.

The uncertainty in length given by the wavelength of light is equivalent to the uncertainty in frequency, given the speed of light is exactly defined. Once the Planck constant (6.62×10^{-34} kg m^2 s^{-1}) is defined, the uncertainty in the electron mass is also determined by the uncertainty in frequency. Therefore, the accuracy in time and frequency measurements is paramount in determining the accuracy associated with all values of physical quantities.

New proposals will also be given in November 2018 to define the unit electric charge (e) and the Boltzmann constant (k_B) as constants without uncertainties. After these definitions, the standards of electric current and temperature will also change from macro-size to micro-size. The electric current of 1 ampere (1 A) is fundamentally the flow of the electric charge of one Coulomb per second. However, a measurement with this defined value at a macro-scale size was not realistic, and the electric current of 1 A has been defined as the 'constant current that produces a force of attraction of 2×10^{-7} newtons (kg m s^{-2}) per meter of length between two straight parallel conductors of infinite length and negligible cross-section placed one meter apart in a vacuum'. After defining the unit of electric charge, the uncertainty of the electric current is also given by those of time and frequency. The product of the absolute temperature and the Boltzmann constant is a parameter associated with the distribution of energy. Therefore, following the definition of the Boltzmann constant, the uncertainty in the temperature is the uncertainty in the energy. Moreover, the uncertainty in the absolute temperature is also given by those of time and frequency (see section 5.6).

References

[1] US Navy 2018 *The Astronomical Almanac Online*
[2] Wikipedia 2018 Orbital period
[3] Vanier J and Audoin C 1989 *The Quantum Physics Atomic Frequency Standards* (Bristol: Adam Hilger), p 610
[4] Heavner T P *et al* 2005 *Metrologia* **42** 411
[5] Chou C W *et al* 2010 *Phys. Rev. Lett.* **104** 070802
[6] Huntemann N *et al* 2016 *Phys. Rev. Lett.* **116** 063001
[7] Ushijima I *et al* 2015 *Nat. Photon.* **9** 185
[8] Nicholson T L *et al* 2015 *Nat. Commun.* **6** 6896
[9] Agnoli P and D'Agostini G 2004 arXiv: physics/0412078v2
[10] Baird K M and Howlett L E 1963 *Appl. Opt.* **2** 455
[11] Michelson A A 1881 *Am. J. Sci.* **22** 120
[12] Brillet A and Hall J L 1979 *Phys. Rev. Lett.* **42** 549
[13] BIPM Resolution of the 1st CGPM1889
[14] Jabbour Z J and Yaniv S L 2001 *J. Res. Natl Inst. Stand. Technol.* **106** 26
[15] NIST *The NIST Reference on Constants, Units, and Uncertainty*
[16] Bartl G *et al* 2011 *Metrologia* **48** S96

IOP Publishing

Measuring Time
Frequency measurements and related developments in physics
Masatoshi Kajita

Chapter 2

Historical role of precision measurements of time and frequency for the development of physics

2.1 Introduction

The concept of time was not a serious issue for our ancestors. Living beings follow their circadian rhythms shaped by sunlight and seasons. There was also no inconvenience to their lives from this lack of accurate information on time. In ancient times, time was more difficult to measure than length and mass.

Compared with our ancestors, modern people are much more preoccupied with time. In the story *Momo*, Michael Ende described this characteristic with the term 'time thief'. In the current era, time and frequency have become physical quantities, which we can measure with the highest accuracy.

This chapter shows the correlation between improvements in time measurements and the development of physics up to the twentieth century.

2.2 Change from Ptolemaic theory to Copernican theory

The concept of time in ancient times was much less serious than now. Around 3000 BC, sundials were used to tell the time of day between sunrise and sunset. A sundial consisted of a flat plate (dial) and a gnomon, which casts a shadow onto the dial (figure 2.1). However, sundials cannot be used at night or during periods of cloudy weather. Usually, they had incorrectly aligned gnomons and hour marks on the dial and hence could not be adjusted to provide the correct time.

Around 1600 BC in Egypt and Babylon, water clocks were also used. There is also a claim that water clocks already existed in China around 4000 BC [2]. Time was measured by the regulated flow of water into (inflow type) or out from (outflow type) a vessel (figure 2.1). The flow rate was assumed to be constant in water clocks but this assumption was not valid because the viscosity of water depends on

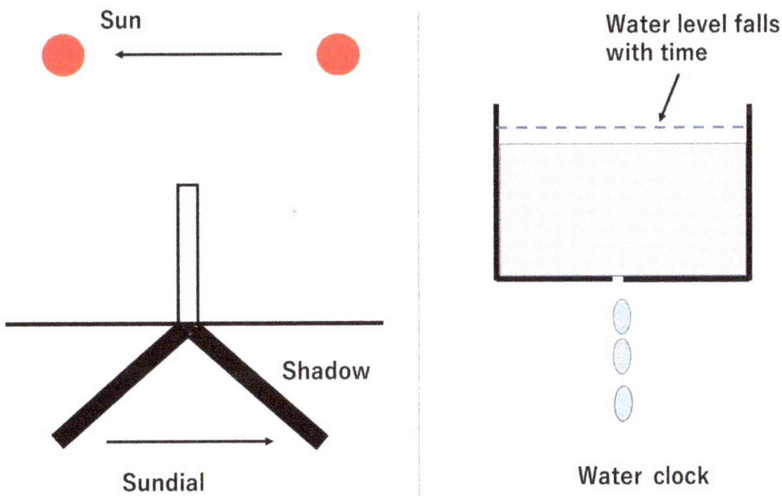

Figure 2.1. Clocks in ancient times: sundial and water clock.

temperature; water at 25 °C is about 25% more viscous than at 30 °C [3]. With changes in temperature of 1 °C, fluctuations of half an hour per day occurred.

Research concerning natural philosophy started in the time of Ancient Greece. Aristotle argued that the Earth was spherical for the simple reason that when going South, constellations along the southern horizon are seen at higher positions, and that at lunar eclipse, we observe a circular shadow of the Earth. However, the Earth was believed to be the center of the Universe, about which the Sun and the stars were believed to revolve. The most famous theory is that of Ptolemy. At that time, the uncertainty in the telling of the time of day was of the order of 1 h, and no discrepancy between observations and Ptolemaic theory were discernible.

Mechanical clocks were developed in medieval Europe. The first system involved gears that rotated by an attached weight falling under gravity. They were made in the high towers of churches to make the length of fall of the weight longer and used to announce the time for prayers with the chiming of bells [4]. These became the standard time-keeping device and their accuracy improved.

With the accuracy of time-keeping improving over the years, discrepancies were realized between Ptolemaic theory and observations for which no reasonable explanation had been given. In 1510, Copernicus was the first to publish his idea that the Earth revolved around the Sun. Called Copernican theory, it was, however, not accepted at that time. This is because he assumed the orbits were circular and his new theory could not account for planetary observations in such detail.

In 1581, Galilei discovered the periodicity of the pendulum's swing, which helped in making rapid improvements in the accuracy of clocks. In 1609, Galilei observed four satellites of Jupiter using a handmade telescope; this provided a step towards establishing Copernican theory. It was finally established between 1609 and 1619

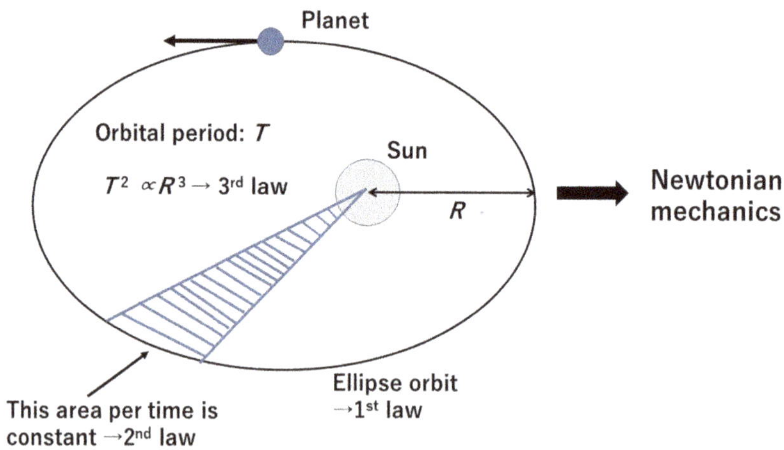

Figure 2.2. Kepler's three laws, on which Newton's theory of universal gravitation was founded.

when Kepler published the following three laws (see also figure 2.2) of planetary motion:

1. The orbit of a planet is an ellipse with the Sun at one of the two foci.
2. A line segment joining a planet and the Sun sweeps out equal areas during equal time intervals.
3. The square of the orbital period of a planet is proportional to the cube of the semi-major axis of orbit.

Kepler presented these laws as simply empirical rules, but he could not explain their appearance. Newton, in his monumental work, derived the three laws of motion (first published in 1687):

1. An object either remains at rest or continues to move at a constant velocity, unless acted on by a force.
2. The sum of the forces F on an object is equal to the mass m of that object multiplied by the acceleration a of the object,

$$F = ma. \tag{2.1}$$

3. When one body exerts a force on a second body, the second body simultaneously exerts a force equal in magnitude and opposite in direction on the first body.

In addition, he presented the law of universal gravitation (published in 1687) showing the gravitational attractive force between two bodies F_G takes the form

$$F_G = G\, Mm/r^2 \tag{2.2}$$

where G is the gravity constant, M and m are the masses of two bodies, and r is the distance between the two bodies. The constant of proportionality between the mass and the gravitational force was derived from the fact that the gravitational acceleration is independent of mass. Newton went on to show that a force inversely proportional to the square of the distance accounts for Kepler's third law. Considering the special case of a circular orbit, with the orbital period denoted by T, the angular velocity ω is given by $(2\pi/T)$ and the centrifugal force F_c experienced by an object with mass m and radius r is given as (at that time there was much debate)

$$F_c = mr\omega^2 = mr(2\pi/T)^2. \tag{2.3}$$

This term is balanced by the gravitational force F_G. Setting $F_G = GMm/r^n$, then

$$\begin{aligned} r(2\pi/T)^2 &= GM/r^n \\ T^2 &= \frac{4\pi^2}{GM}r^{n+1}. \end{aligned} \tag{2.4}$$

To match Kepler's third law, $n = 2$ and hence equation (2.2) is derived. Kepler's second law follows from the conservation of angular momentum $mr^2\omega$. The law of universal gravitation unified the attractive force between planets and stars and the force that makes objects fall to the ground. Moreover, the acceleration due to gravity on the surface of the Earth is almost constant because the height of fall that objects undergo near Earth's surface is much less than the radius of the Earth. While all dynamic phenomena observed at that time were explained by Newtonian mechanics, a mystery remained nonetheless as to why the inertial force and the gravitational force are both proportional to the object's mass. This was to be solved by the theory of general relativity.

In 1656, the pendulum clock was invented by Huygens, and it was the most accurate time-keeper until the 1930s [5]. Note that this occurred roughly in the same period when Newton established his mechanics. This coincidence portends the close correlation between the development of clock technology and physics. Initially, the error in the pendulum clock was 10 min per day but later it was reduced to 1 min per day with better gear-train mechanisms.

The first crystal clock was built in 1927, the accuracy of which was more than one order higher than pendulum clocks and mechanical clocks. It is currently the most widely used time-keeping technology and has remained so despite the development of the atomic clock.

2.3 Measurement of the speed of light

Until the middle of the seventeenth century, no phenomenon had been observed that highlighted the finite propagation time of light. During this era, the speed of light was believed to be either infinite or too fast to measure. However, with the greater accuracy of time measurements, this supposition of the propagation of light was no longer tenable. Indeed, the period taken by Io, one of Jupiter's satellites, to complete an orbit was not constant. In 1676, Rømer considered a finite propagation time for light between Jupiter and Earth. The distance between Earth and Jupiter changes as both planets move around the Sun (figure 2.3). Hence the propagation time of light varies. With this idea, the speed of light was estimated to be 2.2×10^8 m s^{-1}, which is

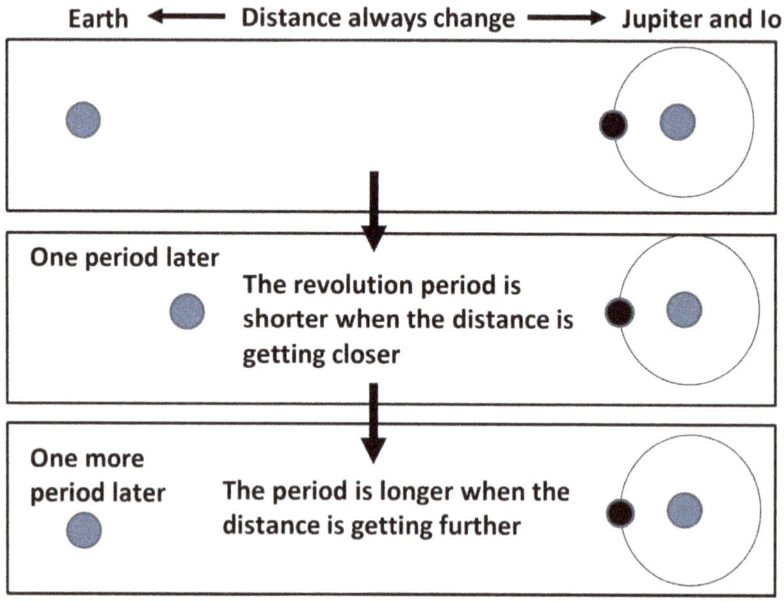

Figure 2.3. The reason Io's orbital period was observed to be irregular.

26% lower than the present value. The irregular orbital period of Io was noted because of improvements in the accuracy of time measurement [6]. However, the distance between Earth and Jupiter was at that time not known to high enough accuracy.

In 1727, Bradley estimated the speed of light from the changes in the observed position of the Draco γ star. With Earth's orbital motion, the direction of light propagation varies with the seasons (figure 2.4). The estimated speed of light was 3.01×10^8 m s^{-1}, which is only a few percentage points different to the present value [7].

The first successful terrestrial measurement of the speed of light was performed by Fizeau in 1849 [8]. With his method, the distance can be measured with high accuracy. Light passing through a gear was reflected by a mirror 9 km away (figure 2.5). The speed of light was determined from the velocity of rotation of the gear for when the reflected light is blocked and was found to be 3.13×10^8 m s^{-1}.

In 1862, Foucault measured the speed of light using the same principle as that used in Fizeau's experiment, except he used a rotating mirror instead of a gear. The measurement gave 2.98×10^8 m s^{-1}, which is 0.6% different from the present value [9]. The improvements to clocks made it possible to determine the rotation velocity with high accuracy.

In the eighteenth to nineteenth centuries, several laws of electricity and magnetism were unified in a set of equations known as Maxwell's equations. These equations indicate that a spatial change in an electric field induces a time-varying magnetic field, and a spatial change in a magnetic field induces a time-varying electric field. The changes in these electromagnetic fields propagate as a wave.

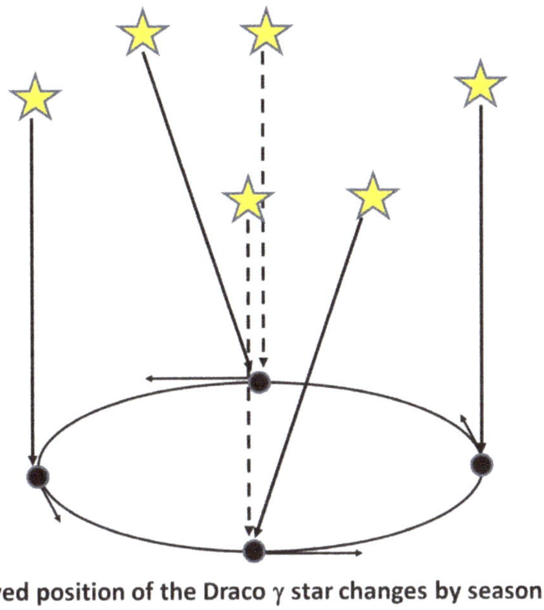

Figure 2.4. The principle of the measurement of the speed of light by Bradley from changes in the observed position of the Draco γ star.

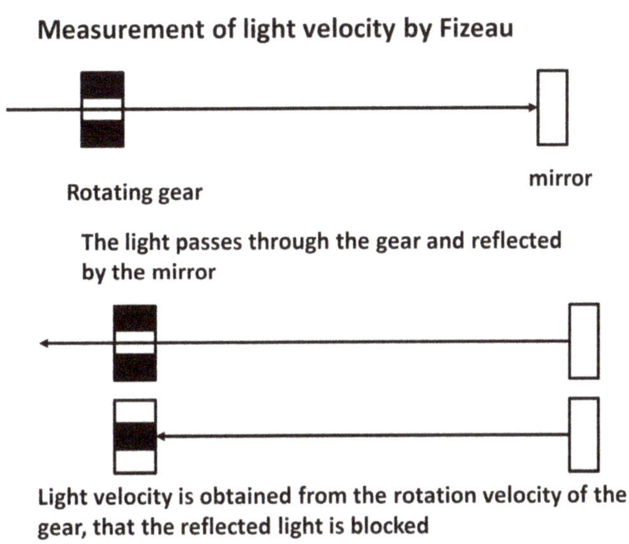

Figure 2.5. The principle involved in the measurement of the speed of light by Fizeau.

Its speed of propagation in a vacuum, as derived by Maxwell's equations, was in good agreement with the speed of light obtained by Fizeau and Foucault. Maxwell fell on the idea that light is an electromagnetic wave. This idea was confirmed by Hertz experimentally in 1887. It was also confirmed that radio and microwaves were also electromagnetic waves with frequencies lower than light. However, a conundrum arose concerning the speed of light in a vacuum.

Generally, velocity depends on the motion of the observer. If observers A and B move along one direction with velocities V_A and V_B, respectively, B will observe the velocity of A to be $V_A - V_B$. However, the speed of light in a vacuum, as given by Maxwell's equations, does not have any dependence on the observer. Hence, the notion of an 'ether' was proposed as the propagator of the light in a vacuum. A different value for the speed of light was expected for observers moving with velocity v within the ether. However, given that the speed of light c was observed by a roundtrip propagation time, the measured velocity should be $c[1 - (v/c)^2]$, therefore, the effect of the motion should appear as a second-order effect. With the velocity of Earth's orbital motion ($10^{-4}c$), the speed of light needs to be measured to an accuracy of 10^{-8} to observe this effect of motion.

Michelson and Morley tried to observe this effect in the speed of light that would arise with Earth's orbital and rotational motion [10]. Using an apparatus known as an interferometer (figure 2.6), light was split by a partial mirror into two waves directed parallel and perpendicular to Earth's motion and reflected by mirrors at distances L_1 and L_2 parallel and perpendicular to the direction of Earth's motion, respectively. The reflected rays of light overlap at the partial mirror again and light interference should be observed related to the difference in the propagation times (T_1 and T_2). If Earth's velocity of motion is v, then

$$T_1 = 2L_1/c[1-(v/c)^2] \tag{2.5}$$

and

$$T_2 = 2L_2/c\sqrt{1 - (v/c)^2}. \tag{2.6}$$

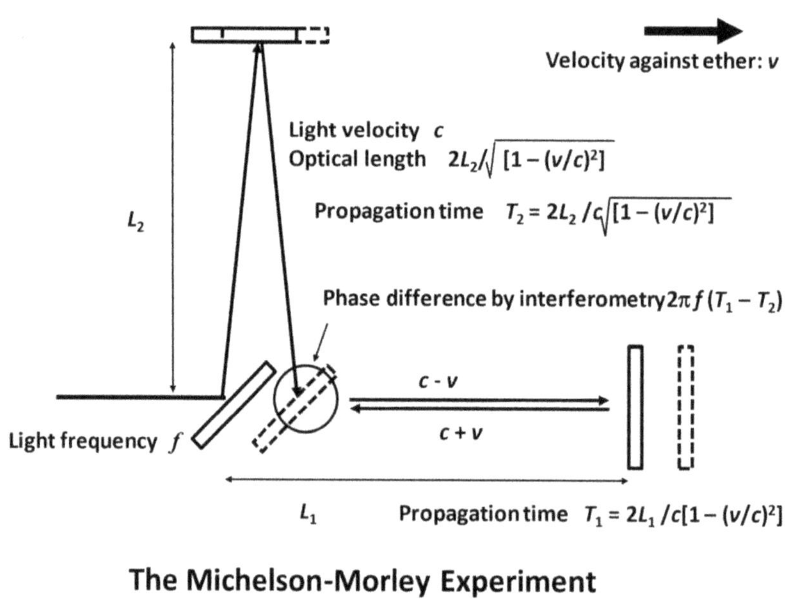

Figure 2.6. Experimental apparatus to look for the influence of Earth's motion on the speed of light.

Earth's velocity of motion changes seasonally with respect to the ether. Assuming $L_1 = L_2 = 9$ m, $T_{1,2}$ are 3×10^{-8} s, it is difficult to measure this difference in $T_{1,2}$ directly. Assuming the frequency of light is $f = 4 \times 10^{14}$ Hz, the difference $(T_1 - T_2)$ $(\delta(T_1 - T_2))$ is 10^{-8} and can be detected by the change in the relative phase $\delta\phi = f\,\delta(T_1 - T_2)$. In this instance, $\delta\phi$ is of order 0.1, which was detectable with the accuracy level of the experiment at that time. No change in the speed of light was observed; indeed, the speed of light was concluded to be constant for all observers with an uncertainty of 10^{-8}. To observe any effect in interferometry, using laser light of a single frequency is preferable. With this setup, Brillet and Hall reduced the experimental uncertainty to 10^{-15} [11].

2.4 Theory of relativity [12]

In 1905, Einstein established the theory of special relativity based on the constant speed of light. When an observer is moving in the x-direction with velocity v, the coordinate of the position and time (x, y, z, t) is observed as (x', y', z', t') which are given by the relations (Lorentz transformation)

$$\mathrm{x}' = \frac{x - vt}{\sqrt{1 - (v/c)^{2,}}}, \; y' = y, \; z' = z, \; t' = \frac{t - vx/c^2}{\sqrt{1-(v/c)^2}}. \tag{2.7}$$

With the same time t',

$$x'_1 - x'_2 = \sqrt{1-(v/c)^2}\,(x_1 - x_2) \tag{2.8}$$

and at the same position x',

$$t'_1 - t'_2 = \sqrt{1-(v/c)^2}\,(t_1 - t_2). \tag{2.9}$$

Therefore, length becomes shorter and the time goes slower in the moving coordinate. The velocity V observed in the moving coordinate system (x', y', z', t') is given by

$$V' = dx'/dt' = (V - v)/(1 - Vv/c^2). \tag{2.10}$$

When $V = c$, $V' = c$ for any value of v.

With Newtonian mechanics, the momentum (p_x, p_y, p_z) is defined by (mv_x, mv_y, mv_z) (m: mass, $v_{x,\,y,\,z}$: velocity in the x, y, z-directions) and a law was established that the total momentum of all bodies is conserved along any direction. With the theory of special relativity, the four-dimensional momentum vector $(p_x, p_y, p_z, E/c^2)$ (E: energy) is defined by

$$p_{x,\,y,\,z} = \frac{mv_{x,\,y,\,z}}{\sqrt{1-(v_x^2 + v_y^2 + v_z^2)/c^2}} \tag{2.11}$$

$$E = \frac{mc^2}{\sqrt{1-(v_x^2 + v_y^2 + v_z^2)/c^2}}. \tag{2.12}$$

The momentum vector is conserved along any coordinate axis. This result shows that the energy becomes infinite when $v = c$. Therefore, a particle with mass cannot be accelerated to a velocity higher than the speed of light. When v is much smaller than c, equation (2.12) is approximated by

$$E = mc^2 + \frac{m(v_x^2 + v_y^2 + v_z^2)}{2}. \tag{2.13}$$

The second term is the kinetic energy defined in the Newtonian mechanics. In contrast to particles with mass, those with zero mass must move with the speed of light. From equations (2.11) and (2.12), the following relation is derived:

$$E^2 = m^2c^4 + c^2(p_x^2 + p_y^2 + p_z^2), \tag{2.14}$$

which is valid for massive and massless particles; the latter includes light, for which the photon has mass zero but nonzero momentum.

With the theory of special relativity, Maxwell's equations for electromagnetism are valid in all coordinate systems. The Lorentz transformation is performed with four-dimensional electromagnetic potential (A_x, A_y, A_z, V_e/c^2), where $A_{x,\ y,\ z}$ is the magnetic field vector potential and V_e is the electric potential. The electric and magnetic field strengths depend on the observer's frame of reference. We consider an electric current as viewed by observer A as having a positive electric charge of density $+\rho$ moving with velocity $+V$ and a negative electric charge with density $-\rho$ moving with velocity $-V$. Although the electric field is zero, there is a magnetic field induced by the current of $2\rho V$. For an observer B co-moving with the positive charge, the velocity of the negative charge is not $2V$, but $2V/(1 + V^2/c^2)$. Hence the magnetic field for observer B is smaller than that for observer A. Moreover, the unit of length for the negative charge is shorter, then the density of the negative charge is higher than that of the positive charge and the total electric charge density is $\rho[1 - 1/\sqrt{1 - (2V/c)^2}]$. Therefore the electric field is non-zero.

The theory of special relativity shows that Newtonian mechanics is valid only in an approximation when the velocities involved are much lower than the speed of light. In addition, the theory of special relativity is valid only for the inertial system (a system without acceleration).

In 1916, Einstein published the theory of general relativity, based on the 'equivalence principle'. In Newtonian mechanics, a mystery arose as to why the inertial force of an object and the gravitational force acting on it are both proportional to its mass. The equivalence principle considers that both forces are one and the same. When we are in an accelerated system, we see that the optical path is bent (figure 2.7). Einstein expected that the bending of the optical path is also caused by gravity. Then the position of stars should be observed at a shifted position when the optical path taken by light passes by a large nearby massive star. This expectation was confirmed in 1919 by Eddington when observing a distant star during a solar eclipse [13].

The length of the bent optical path is longer than the straight one. As the speed of light is an absolute constant, the propagation time is longer. Therefore, time goes

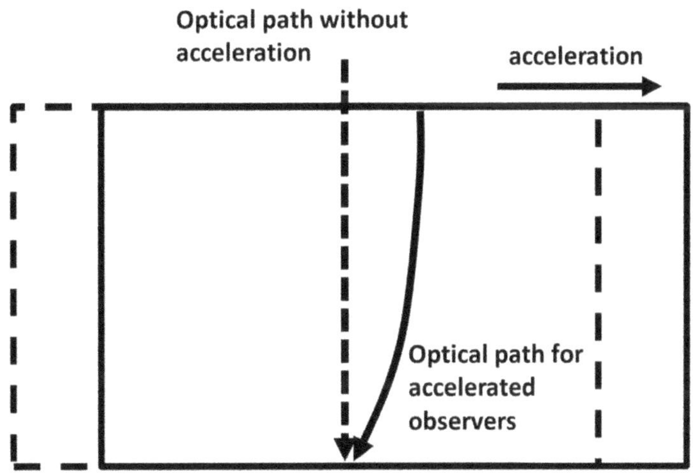

Figure 2.7. Bending of an optical path in an accelerated frame of reference. Einstein considered the same phenomenon also with gravity.

slower in a gravitational field (gravitational redshift). It is not simple to obtain the formula for the redshift observed given the gravitational potential $m\Phi$ (m: mass), but it is approximately given by

$$t'_1 - t'_2 = \sqrt{1-(2\Phi/c^2)}\,(t_1 - t_2). \tag{2.15}$$

We can understand this formula using the theory of special relativity with $mv^2 = 2m\Phi$ from the change in kinetic energy induced by the acceleration due to gravity.

Within Newtonian mechanics, gravity is an attractive force between matter having mass. However, gravity also bends optical paths despite light having no mass. With the theory of general relativity, gravity distorts the space (bending the coordinate axis) and matter moves uniformly along the bent axis. The distortion of space propagates as a wave (a gravitational wave) with the speed of light. We can imagine a model with the object moving on a flexing mat (figure 2.8). The search for gravitational waves was one of the most active subjects for researchers and was recently successful with the first detection in 2015 (see section 4.2.2).

Exercises

2.1 For any satellite, the relation between the energy in orbital motion (K) and the gravitational potential energy (P_G) are given by $P_G = -2K$. Derive this relation.

2.2 Consider an object moving with velocity V in the x-direction. With the theory of special relativity, the momentum along the x-direction p_x and energy E of the object are defined by

$$p_x = \frac{mV}{\sqrt{1-(V/c)^2}} \qquad E = \frac{mc^2}{\sqrt{1-(V/c)^2}}.$$

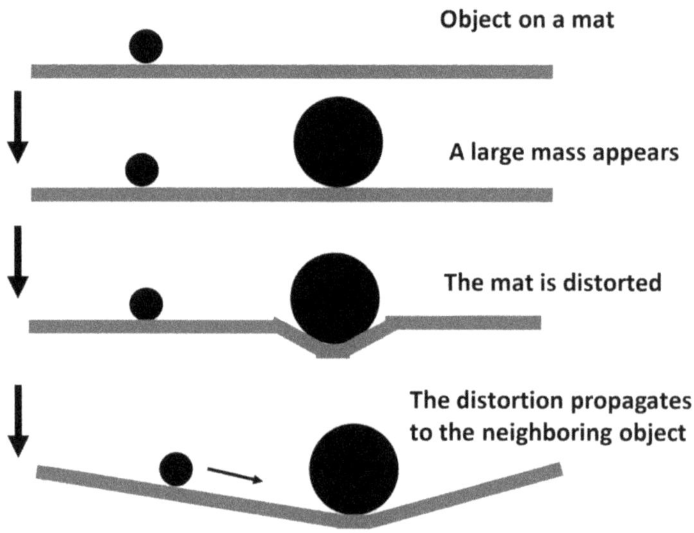

Figure 2.8. Space distortion induced by gravity can be imagined with the distortion of the surface of a flexing mat when a large object is placed on it.

Using equation (2.10), derive both quantities for another observer moving in the x-direction. With this calculation, we can confirm that the Lorentz transformation is valid also for the momentum vector.

Solutions

2.1 We denote the orbital radius of a satellite moving around the Earth by R. The masses of the Earth and the satellite are denoted by M and m, respectively. The gravitational force is given by

$$F_G = \frac{GMm}{R^2}$$

and the gravitational potential is

$$P_G = -\frac{GMm}{R}.$$

The gravitational force must balance with centrifugal force (angular velocity ω) given by $F_c = mR\,\omega^2$. Taking

$$F_G = F_c, \quad m(R\Omega)^2 = \frac{GMm}{R} = -P_G.$$

The velocity of the satellite is

$$v = R\Omega \text{ and } K = \frac{mv^2}{2} = \frac{m(R\Omega)^2}{2}.$$

Then $P_G = -2K$ is derived.

2.2 For the observer moving with velocity v, the velocity of the object is given by

$$V' = (V - v)/[1 - Vv/c^2].$$

Then

$$p_x' = \frac{m(V - v)}{(1 - Vv/c^2)\sqrt{1 - ((V - v)/(1 - Vv/c^2))^2/c^2}}$$

$$E' = \frac{mc^2}{\sqrt{1 - ((V - v)/(1 - Vv/c^2))^2/c^2}}.$$

From

$$(1 - Vv/c^2)^2[1 - ((V - v)/(1 - Vv/c^2))^2/c^2] = (1-(V/c)^2)(1-(v/c)^2)$$

$$p_x' = \frac{m(V - v)}{\sqrt{1-(V/c)^2}\sqrt{1-(v/c)^2}} = \frac{p_x - vE/c^2}{\sqrt{1-(v/c)^2}}$$

$$E'/c^2 = \frac{E/c^2 - vp_x/c^2}{\sqrt{1-(v/c)^2}}.$$

Therefore, the Lorentz transformation is also valid for the four-dimensional vector of momentum.

References

[1] Moss T 2013 *How Do Sundials Work?* (London: British Sundial Society)
[2] Cowan H J 1958 *Time and Its Measurement From the Stone Age to the Nuclear Age* (Cleveland, OH: The World)
[3] Goodenow J, Orr R and Ross D 2007 *Mathematical Model of Water Clocks* (Rochester, NY: Rochester Institute of Technology)
[4] Usher A P 1988 *A History of Mechanical Inventions* (New York: McGraw-Hill), p 194
[5] Milham W I 1945 *Time and Timekeeping* (New York: MacMillan), p 330
[6] Filonovich S R 1986 *The Greatest Speed* (Moscow: Mir), p 285
[7] Bradley J 1728 *Philos. Trans. R. Soc. Lond.* **35** 637
[8] Fizeau H 1849 *C. R. Hebd. Seances Acad. Sci.* **29** 90
[9] Foucault L 1862 *C. R. Hebd. Seances Acad. Sci.* **55** 501
[10] Michelson A A 1881 *Am. J. Sci.* **22** 120
[11] Brillet A and Hall J L 1979 *Phys. Rev. Lett.* **42** 549
[12] Einstein A 1916 *Relativity: The Special and General Theory* (New York: H Holt)
[13] Dyson F W *et al* 1920 *Philos. Trans. R. Soc.* A **220** 291

Chapter 3

The fundamentals of an atomic clock

3.1 Introduction

At the beginning of the twentieth century, quantum mechanics was established, showing that:

(1) Electromagnetic waves (such as light, microwaves, ultra-violet radiation) have characteristics of particles (called photons) having energy $E = hf$ (h: Planck constant, f: frequency) and momentum $p = h/\lambda$ (λ: wavelength).

(2) All particles have characteristics of waves. Their energy is given by $E = hf$ and their momentum by $p = h/\lambda$ (the same as for electromagnetic waves). As there is no significant change in the amplitude of the wavefunction within the phase difference of ± 1 (the phase has no physical appearance), there is an uncertainty principle between 'time and energy' and 'position and momentum'. When the particle is localized in an area of size L, the wavelength can only assume discrete values $\lambda = 2L/n$ (n—integer), and hence also the energy and momentum of the particles. The waveform is obtained from the relation between the frequency and the wavenumber in accordance with the relation between E and p obtained either by applying Newtonian mechanics (the Schrödinger equation) or theory of special relativity (the Dirac equation). From the Dirac equation, the phase propagation speed is c for massless particles but is faster than c for particles with mass (the theory of special relativity is not violated because the motion of matter is given by the group velocity, which is less than c).

(3) The energy of the electrons in an atomic state can be changed ($E_1 \rightarrow E_2$) by absorbing or emitting a photon with an energy equal to the energy difference in the initial and final states (atomic transition energy). The total energy before and after the transition must be conserved. As stated above, the energy of an electron in an atom can only take discrete values.

doi:10.1088/978-0-7503-2124-2ch3

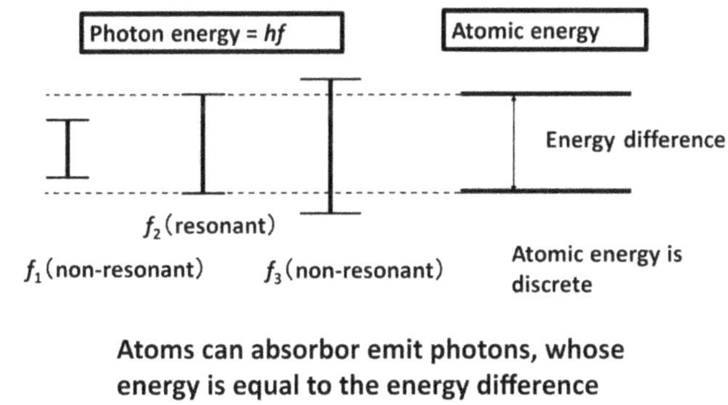

Figure 3.1. Fundamentals of an atomic clock based on the atomic transition frequency.

Therefore, atoms can only absorb or emit electromagnetic waves of a specific frequency satisfying $f = |E_1 - E_2|/h$, known as the transition frequency or resonance frequency (figure 3.1).

An atomic clock operates based on a specific frequency of the electromagnetic wave absorbed or emitted by an atomic or molecular system. The first clock was developed in 1949 using ammonia molecules, although the improvement in frequency was not so dramatic in comparison with the quartz clock, which operates based on the regular vibrations of a quartz crystal resonator in the shape of a tuning fork. In 1955, an atomic clock based on the cesium (Cs) atom was constructed at the National Physical Laboratory (UK); at that time, the frequency accuracy was 10^{-10} (a 1 s error per 300 years). The accuracy of atomic clocks was further improved; details to be given in sections 3.2 and 3.3. The improvement is achieved for the following reasons:

(i) The atomic transition frequency is discrete, in that with a slight shift in the frequency of the electromagnetic wave, the atomic system becomes non-resonant.

(ii) In a gaseous state, the atoms are isolated from one another. The characteristics of a single atom are therefore uniform over the gaseous state. In contrast, the bonding state between neighboring atoms in a solid substance cannot be perfectly the same throughout the solid and depends on the circumstances (temperature, composition, impurities).

(iii) The atomic energy structure is determined mainly by the Coulomb force between the nucleus and the electrons because at the atomic scale the electric field is more than five orders larger than the electric fields that we can generate in the laboratory.

However, even atomic clocks cannot have infinite accuracy because the advantages stated above are not perfectly satisfied. In section 3.2, the attainable accuracy with atomic clocks is discussed. In section 3.3, we also discuss the notion of 'accuracy' and 'stability' and elaborate on the performance of atomic clocks as determined by their uncertainties.

3.2 Attainable accuracy of atomic clocks

The uncertainty in time and frequency readings obtained from an atomic clock are given in terms of the statistical uncertainty and the systematic uncertainty (figure 3.2). In this chapter, both types of uncertainties are explained.

3.2.1 Statistical uncertainty

A certain width in the frequency of a resonance (linewidth) exists for any atomic or molecular transition. The transition frequency is measured using the phase estimation procedure for ϕ in a finite measurement time τ, where $\phi = 2\pi f\tau$ (figure 3.3). The measurement time is limited by the interaction time between the atom and the electromagnetic wave. However, it can also be limited by a phase jump, by an interatomic collision, or a spontaneous emission (a transition to a lower energy state emitting a photon in a random direction). Therefore, ϕ is unknown up to a complete cycle, and therefore, assuming a random probability distribution of the possible outcomes of the phase, f has an uncertainty given by

$$\Delta f = 1/2\pi\tau, \tag{3.1}$$

which is observed as a linewidth. This linewidth obeys the uncertainty principle between energy and time, which is one of the most important principles in quantum mechanics.

The measured values of frequency f_{obs} are randomly distributed within $f \pm \Delta f$ and the real value of the transition frequency f is obtained from the average of the measurement results. The statistical measurement uncertainty with the sample number N_S is given by

$$\Delta f_{\text{static}} = \Delta f / \sqrt{N_S}. \tag{3.2}$$

The sample number is given by N_A (τ_e/T), where N_A is the number of atoms present during the measurement, τ_e the period for a single measurement cycle, and T the measurement time. Therefore,

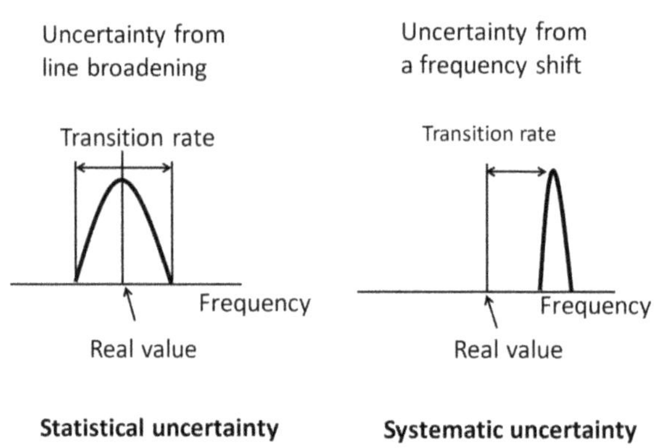

Figure 3.2. Illustration of statistical and systematic uncertainties.

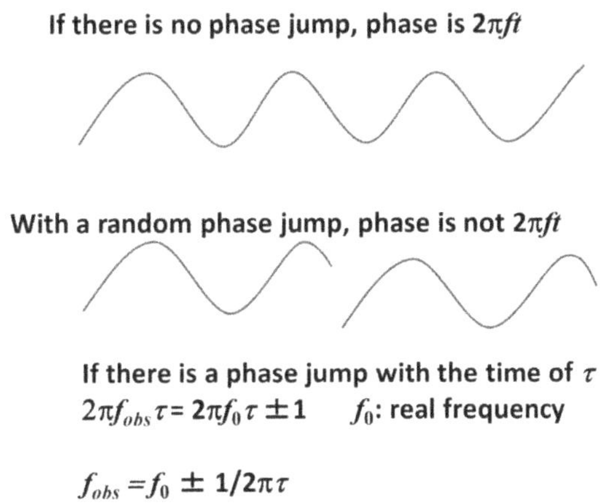

Figure 3.3. Fundamentals of line broadening given by a phase jump.

$$\Delta f_{\text{statistic}} = \Delta f \sqrt{\tau_e / N_A T}. \tag{3.3}$$

The statistical uncertainty is reduced by measuring many atoms over long measurement times. Note also $\tau_e > 1/2\pi\Delta f$. When $\tau_e \approx 1/2\pi\Delta f$, the statistical uncertainty is proportional to $\sqrt{\Delta f}$.

3.2.2 Systematic uncertainty

Depending on the circumstances, shifts in the measured atomic transition frequency also exist. For commercial atomic clocks, these shifts create an uncertainty in the measured time and frequency. With precision atomic clocks developed in the laboratories (called primary frequency standards), we can estimate the shifts induced by several causes and give corrections. In such instances, the uncertainty associated with the shifts generates uncertainty in the measurement. This uncertainty given by the frequency shift is called 'systematic uncertainty'. In this chapter, the causes of the frequency shifts are itemized and discussed.

To begin, the frequency shift induced by the atomic motion is described. When atoms are moving with velocity v_x parallel to the propagation of the electromagnetic wave (velocity c), the atomic transition frequency of f is measured to be

$$f_{\text{obs}} = f[1-(v_x/c)]. \tag{3.4}$$

Called the first-order Doppler shift, this shift is not problematic when the average of v_x is zero. The first-order Doppler shift is seen as a broadening in the transition frequency distribution measured using atoms in free space. This Doppler broadening is eliminated by several methods, for example, measuring localized atoms in an area smaller than the wavelength of the electromagnetic wave.

As described in section 2.4, time goes slower in a moving body. The transition frequency of atoms (real transition frequency f) moving with velocity v (much less than c) is measured as

$$f_{\text{obs}} = f\sqrt{1-(v/c)^2} \approx f[1-v^2/2c^2]. \tag{3.5}$$

This frequency shift is called the second-order Doppler shift, which for a given v is much smaller than the first-order Doppler shift but is independent of the direction of motion; its average is nonzero. Therefore, the second-order Doppler shift is much more problematic than the first-order Doppler shift. The velocity of atoms in the gaseous state is of order 200–400 m s^{-1} at room temperature. The ratio of the second-order Doppler shift to the transition frequency is of order -10^{-12}. By measuring the velocity distribution, the average of the second-order Doppler shift is estimated and corrected; the fractional uncertainty can be reduced to below 10^{-15}.

In section 2.4, time was also shown to slow down in a gravitational field (gravitational redshift). In taking measurements at a height 1 m higher, the measured transition frequencies of atoms are higher by a factor 10^{-16}. Therefore, we have to stipulate in definition that the fundamental standard is that value obtained from a measurement at the geoid surface. Corrections are given based on the altitude of the location where the measurement had taken place.

Electrical effects give significant frequency shifts. The electric energy shift of a state a, denoted $\Delta E_e(a)$, is given by

$$\Delta E_e(a) = \int \rho(\mathbf{r}) V_e(\mathbf{r}) d\mathbf{r}, \tag{3.6}$$

where $V_e(\mathbf{r})$ and $\rho(\mathbf{r})$ are the electric potential and the electric charge density, respectively. Taking the origin of the coordinate at the center of mass of atoms or molecules, then Taylor series expansion yields

$$V_e(\mathbf{r}) = V_e(0) + \frac{dV_e}{d\mathbf{r}}\mathbf{r} + \frac{1}{2}\frac{d^2V_e}{d\mathbf{r}^2}\mathbf{r}^2 = V_e(0) - E_L\mathbf{r} - \frac{1}{2}\frac{dE_L}{d\mathbf{r}}\mathbf{r}^2, \tag{3.7}$$

where E_L is the electric field strength. Substituting this expansion into equation (3.6) yields various contributions to this shift

$$\Delta E_e = \Delta E_C + \Delta E_S + \Delta E_Q,$$

defined as

$$\begin{aligned}
\Delta E_C &= V_e \int \rho d\mathbf{r} = V_e q, && \text{Coulomb energy shift} \\
& && (q\text{: total electric charge}) \\
\Delta E_S &= -E_L \int \rho\mathbf{r}\, d\mathbf{r} = -dE_L, && \text{Stark energy shift} \\
& && (d\text{: electric dipole moment}) \\
\Delta E_Q &= -\frac{dE_L}{d\mathbf{r}} \int \frac{\rho \mathbf{r}^2}{2}\, d\mathbf{r} = -\frac{dE_L}{d\mathbf{r}} Q, && \text{Electric quadrupole energy shift} \\
& && (Q\text{: electric quadrupole moment}).
\end{aligned} \tag{3.8}$$

The shift in the transition frequency for transition a–b is then given by

$$\Delta f_e = \frac{\Delta E_e(b) - \Delta E_e(a)}{h}. \tag{3.9}$$

The Coulomb energy shift (zero for neutral atoms and molecules) does not contribute to the shift in the transition frequency because it is exactly the same for all energy states. The Stark shift depends on the energy state, and it gives a significant shift in the transition frequency. The electric dipole moment d is zero for atoms and non-polar molecules in a field-free space. With a small electric field, d is proportional to E_L and the Stark shift is proportional to $E_L{}^2$ (second-order Stark shift). The second-order Stark shift is induced by electromagnetic waves irradiated from the probe or through manipulations. Sometimes the second-order Stark shifts induced by blackbody radiation are significant. In addition, the electric quadrupole shift has a dependence on the energy state, but its contribution to the transition frequency is much less than that for the Stark shift, except when the electric field gradients are large.

There is also an energy shift induced by the interaction between the magnetic dipole moment μ_m and the magnetic field B, which is called the Zeeman energy shift. This Zeeman shift is given by

$$\begin{aligned} \Delta E_Z &= \mu_B B(g_L M_L + g_S M_S + g_I M_I + g_R M_R) \\ g_L &= 1, \ \ g_S = 2.002\ 319, \ \ g_I < 10^{-3}, \ \ g_r < 10^{-4} \end{aligned} \tag{3.10}$$

where μ_B is the Bohr magneton (μ_B/h = 1.4 MHz G^{-1}), g the g-factor, and M the component of angular momentums parallel to the magnetic field. Subscripts L, S, I, and R denote the electron orbital angular momentum, electron spin, nuclear spin, and molecular rotation, respectively. We imagine the particles with spin (electron and nucleus) to behave like permanent magnets. The dipole magnet moment induced by the orbital motion is similar to that for an electromagnet. The wavefunction of each energy structure Φ is given by

$$\Phi = \sum p_i \langle M_L^i,\ M_S^i,\ M_I^i,\ M_R^i \rangle,\ M_L^i + M_S^i + M_I^i + M_R^i = M\ (\text{constant}) \tag{3.11}$$

and

$$M_{L,S,I,R} = \sum |p_i|^2 M_{L,S,I,R}^i. \tag{3.12}$$

The Zeeman shift is generally nonlinear because $_{pi}$ depends on the magnetic field. For simplicity, an example of the hydrogen (H) atom in the $1^2S_{1/2}$ (F, M) state is presented; here, the electron spin and H-nuclear spin are 1/2, and F denotes the hyperfine structure. The hyperfine structure gives a slight energy difference induced by the magnetic interaction between the electron and nucleus. When both spins are parallel ($F = 1$), the magnetic force is repulsive, and when anti-parallel ($F = 0$), it is attractive. The $(F, M) = (1,\pm1)$ states are given by $\langle M_S, M_I\rangle = \langle \pm1/2, \pm1/2\rangle$ and the Zeeman shift is strictly linear with the coefficient of proportionality being $\pm$1.4 MHz G^{-1}. The (F, M) = (1,0) and (0,0) states are given by $p_1\langle 1/2, -1/2\rangle + p_2\langle -1/2, 1/2\rangle$ and $p_2\langle 1/2, -1/2\rangle - p_1\langle -1/2, 1/2\rangle$, respectively. In the absence of a magnetic field, $p_1 = p_2 = 1/\sqrt{2}$, and in the limit of high magnetic field the coefficients converge to $p_1 = 1$ and $p_2 = 0$. For small magnetic field strengths, the Zeeman shift is approximately given by the formula

$$\Delta E_Z = C_1 MB + C_2(M)B^2 \quad \text{where} \quad C_2(M) = C_2(-M). \tag{3.13}$$

The first-order Zeeman shift is zero with $M = 0$. For the states with $C_2(M) = 0$, the Zeeman energy shift is perfectly eliminated when averaged over the energies in the $\pm M$ states. The Zeeman shift is suppressed for transitions with $\Delta M_L = \Delta M_S = 0$. To minimize the Stark and Zeeman effects, the following methods are used:

1. Measure the transition frequency where the Stark and Zeeman shifts are relatively small. There is also a possibility that the energy shift for each state is large but the shifts for the two states of the transition are equal and opposite, and hence cancel each other.
2. Another transition frequency with a large shift is also measured, from which the electromagnetic field is estimated. Next, the shift in the standard transition frequency is estimated and corrected.
3. The difference between different $M = M'' \rightarrow M'$ transition frequencies is induced by the Stark, Zeeman, and electric quadrupole shifts (sub-components of the transition). The first-order terms of these shifts are eliminated by averaging the transition frequencies of sub-components. For example, the first-order Zeeman shift is eliminated by averaging the transition frequencies of $M = M'' \rightarrow M'$ and of $-M'' \rightarrow -M'$, as shown above.

In some cases, collisions between atoms and molecules can also cause shifts in transition frequencies (collision shifts). Such shifts are proportional to the density of atoms and molecules. Therefore, the dependence on the density of atoms or molecules is introduced by a factor multiplying the expression for the transition frequency assuming no collisions.

3.3 Accuracy and stability

The performance of atomic clocks is discussed with regard to accuracy and stability. Figure 3.4 shows the results of distributions of measurements with regard to 'high' and 'low' levels of 'accuracy' and 'stability'. Accuracy indicates the reliability of the final results after averaging many experimental results and giving corrections to frequency shifts induced by the different transitions. The accuracy can also be high if the difference between each measurement results is large. To achieve high accuracy, the systematic uncertainty should be small. The statistical uncertainty can be reduced by taking measurements over long time periods.

Stability refers to the constancy of each measurement result. The stability can also be high if the measured frequencies are equally shifted from the defined frequency. With time-averaging arises the consideration of 'short-term stability' and 'long-term stability'. The former is determined from the low statistical uncertainty covering short measurement times. To achieve high 'short-term stability', high Q-values (defined as transition frequency/spectrum linewidth) and large numbers of atoms or molecules are required. Taking long measurement times, the statistical uncertainty is

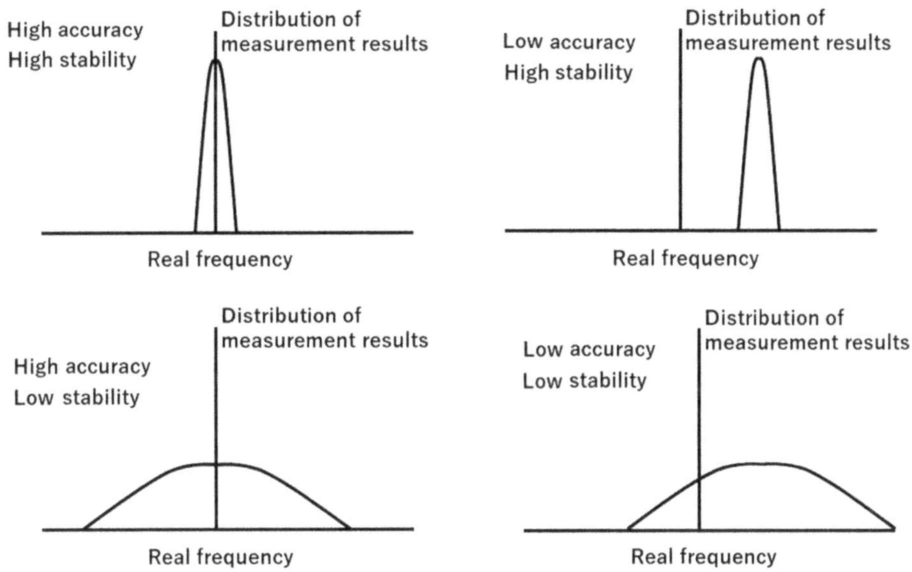

Figure 3.4. Schematic of results for the distribution of measurements with high and low 'accuracy' and 'stability'.

suppressed, but the influence of the change in conditions (e.g. electric field and magnetic field) becomes significant. Therefore, 'long-term stability' is determined from the systematic uncertainty.

3.4 Atomic clock operating in the microwave region

The first atomic clock, constructed in 1955, used microwaves because microwave frequency could be measured with counters. The operation of these clocks was based on the $^2S_{1/2}$ $F = 3$–4 transition (F: hyperfine structure) in cesium (Cs). This frequency was defined to be the standard for time and frequency measurements in 1967. Why was the Cs atomic transition chosen as the standard? The Cs atom is a stable alkali atom (with a single valence electron, the energy structure is therefore relatively simple) with the largest mass (the Fr atom is radioactive). With larger mass, the second-order Doppler shift is smaller. Moreover, the transition frequency in the microwave region is highest for the alkali atoms, and hence the time unit of 1 s can have a finer scale division. Therefore, the atomic transition frequency of Cs was best suited to achieve the highest accuracy [1, 2].

The microwave transition is observed with the change in population of the two energy states. However, the energy difference between these two states is much less than the kinetic energy of the atoms. Therefore, the natural population of the states is an almost-equal balance. To observe a change in population of each state induced by the transition, the initial atomic state is required to be prepared in one of the two states.

The structure of an atomic clock using a beam of Cs atoms is illustrated in figure 3.5 [3]. First, an oven produces an atomic beam that passes through a magnetic

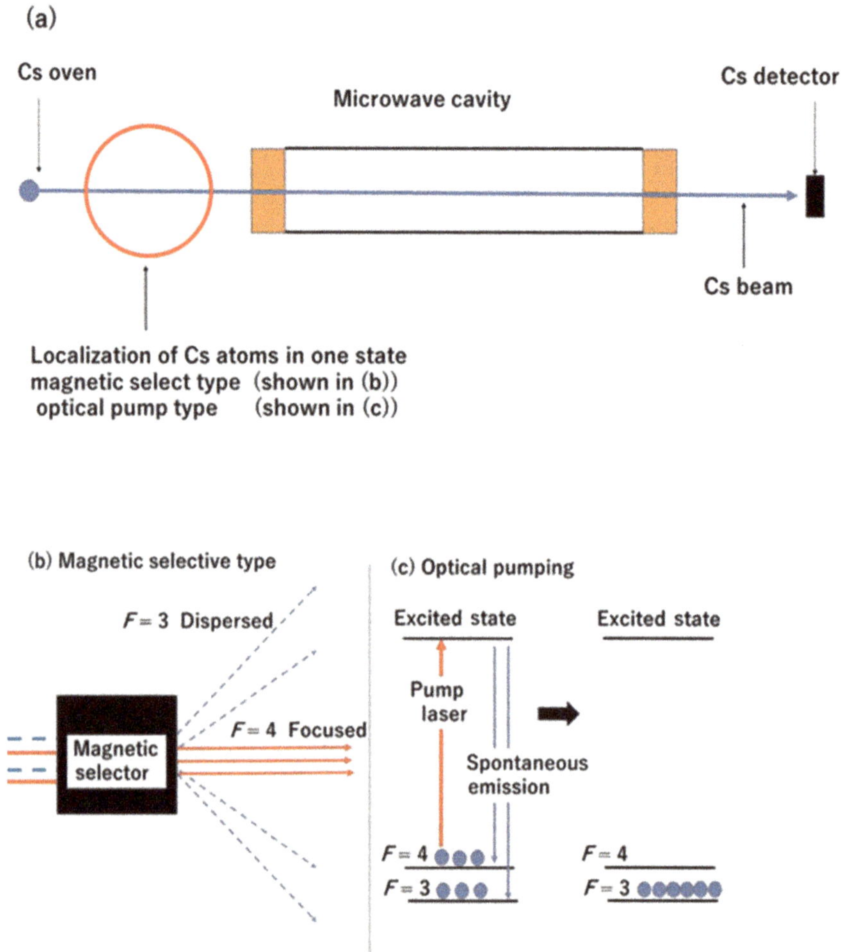

Figure 3.5. Schematic of the structure of a Cs atomic clock employing an atomic beam: (a) the whole apparatus including the system that localizes atoms in one of two hyperfine states ($F = 3, 4$); (b) the system of magnetic selector; and (c) diagram depicting the mechanism of optical pumping.

selector, which is designed to focus atoms of Cs in the $F = 4$ state and disperse those in the $F = 3$ state. The atomic beam with localized $F = 4$ Cs atoms passes through a microwave cavity. The length of the microwave cavity is roughly 150 cm for laboratory-type (commercial-type cavities are 20 cm long). The atoms pass through the cavity in about 0.005 s achieving a linewidth of about 50 Hz. After interacting with the microwave, atoms in the $F = 3$ state are removed by the magnetic selector; the population in the $F = 4$ state is then monitored. When the microwave is tuned to the transition, the population in the $F = 4$ state after interaction is small.

Another method to localize the atomic state is optical pumping—irradiating laser light that is tuned to the transition from one of two states to a higher energy state. From the excited state, spontaneous emission takes place to repopulate the two lower energy states. Repeated excitation and emission transfers atoms in one state to

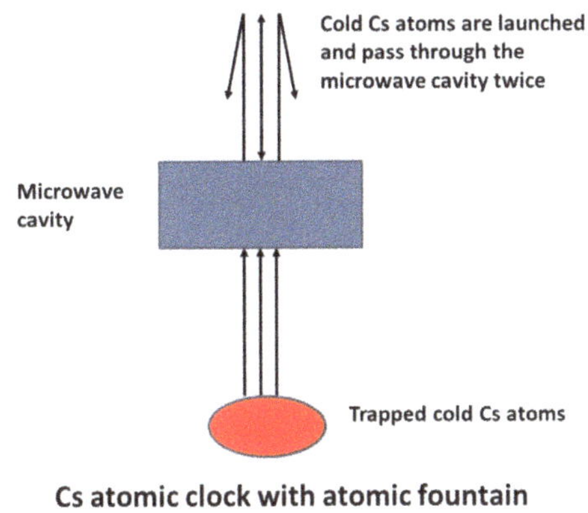

Figure 3.6. Structure of a Cs atomic clock with an atomic fountain.

the other. With this method, a signal intensity higher than the magnetic selective type is achieved. The microwave transition rate is monitored from the fluorescence induced by the laser light, which is tuned to the transition of Cs in the $F = 4$ state.

The Cs atomic clock is based on the $^2S_{1/2}$ (F, M) = (3,0)–(4,0) transition frequency. This transition is used because there is no first-order Zeeman shift. A bias magnetic field is applied to avoid overlaps with other sub-component transition lines. Therefore, there is a second-order Zeeman shift of order 10^{-10}, which is corrected by measuring the magnetic field from the first-order Zeeman shift in the transition frequencies of other sub-components. The second-order Doppler shift is of order 10^{-12}, which is corrected measuring the velocity distribution in the atomic beam. Measurements with uncertainties of 7×10^{-15} have been reported using optical pumping [4], while it was 10^{-14} using a magnetic selector.

Using slow atoms, the interaction time between the atom and microwave is longer and the linewidth becomes narrower. The second-order Doppler shift is also reduced. The development of laser cooling technology (see appendix A) allowed the velocity of the Cs atoms to be lowered to about a few cm s^{-1}. In 1991, the first fountain-type Cs atomic clock (figure 3.6) was developed at the Systèmes de Référence Temps Espace, France [5]. Atoms that have been laser cooled and trapped are launched upward using laser light. The launched atoms pass through a microwave cavity, reverse direction and fall through the cavity again. The microwave transition rate is monitored after the second passage through the cavity from fluorescence induced by laser light tuned to the transition from the $F = 4$ state. The resonance linewidth, given by the period between the first and the second passage through the cavity, is 1–2 Hz. With this narrow linewidth, overlaps from the sub-component transition lines are avoided by using a bias magnetic field weaker than that used for the atomic beam. A second-order Zeeman shift of less than 10^{-16} is produced; the second-order Doppler shift is less than 10^{-17}. The collision shift is significant for the atomic fountain apparatus and a correction is obtained from the dependence of the measured frequency on the density of launched atoms. Cs atomic clocks using an atomic fountain were also constructed in the US, Germany, the UK, Italy, Japan, and China. A fractional uncertainty of order 10^{-16} was obtained from several groups [6].

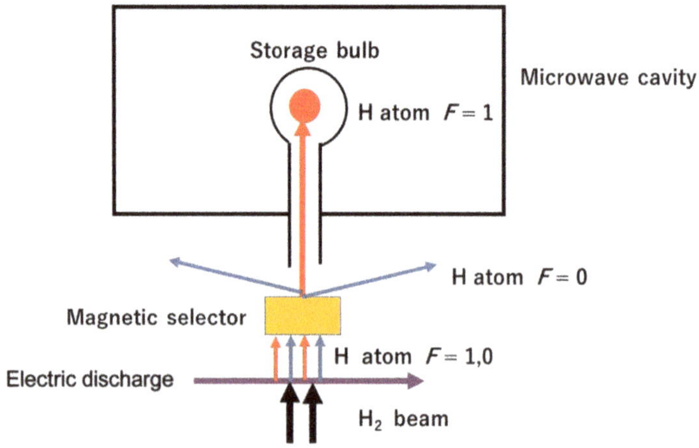

Figure 3.7. Structure of a hydrogen maser. H atoms in the $F = 1$ state are selectively guided to the storage bulb.

Apart from Cs clocks, several other atomic clocks that also operate in the microwave region were developed. The hydrogen maser is based on the hyperfine transition ($^2S_{1/2}$ $F = 0 \rightarrow 1$) of the hydrogen (H) atom (frequency 1.4 GHz) [7]. H atoms are produced by the dissociation of H_2 molecules using electric discharge and the state is selected by a magnetic selector. Atoms in the $F = 1$ state are guided to a storage bulb, the interior wall of which is coated by Teflon (see figure 3.7). The storage bulb is in turn inside a microwave cavity tuned to the transition frequency. The microwave mode within the cavity is amplified by stimulated emission from the H atoms in the $F = 1$ state.

During collisions with the Teflon-coated wall, there is no phase jump (see section 3.2.1) induced for the hyperfine transition of the H atom; therefore, the observed linewidth is narrow. The first-order Doppler effect is negligibly small because the H atoms are in the storage bulb, which is smaller than the microwave wavelength. We can observe low statistical uncertainty with short measurement times (10^{-15} with a measurement time of 1000 s). However, collisions with the wall induce a frequency shift that drifts as time elapses because the chemical structure of Teflon changes. Therefore, rather than being an accurate frequency source, the hydrogen maser is useful as a frequency source oscillator with high stability covering short measurement times. When frequency accuracy is required, corrections by other atomic clocks with smaller systematic uncertainty are performed.

A ^{87}Rb atomic clock based on the $^2S_{1/2}$ $F = 1 \rightarrow 2$ transition (6.8 GHz) was also developed [8]. This clock is more compact than the Cs atomic clock as the localization of atoms to the $F = 2$ state is performed with atoms in a cell using ^{87}Rb lamp filtered using a ^{85}Rb cell (figure 3.8). The ^{87}Rb lamp emits light of two components oscillating at the transition frequency corresponding to the excited state to the $F = 1$ and $F = 2$ states. The component related to the $F = 2$ state is absorbed by the ^{85}Rb filter, being resonant with the transition from the ^{85}Rb $F = 3$ state. The other component, related to the transition from the $F = 1$ state, irradiates other ^{87}Rb

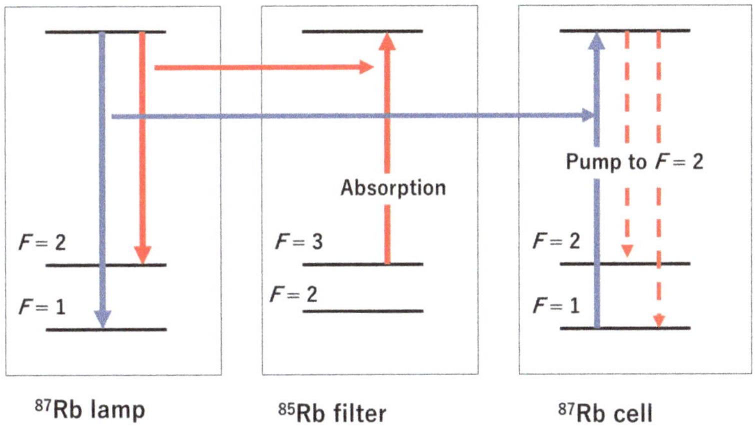

Figure 3.8. Schematic of optical pumping by a Rb lamp.

atoms in the cell and pumps the ^{87}Rb atoms to the $F = 2$ state. Optical pumping was not possible for Cs atoms until the laser diode was developed.

A ^{87}Rb atomic clock with an atomic fountain was also developed with which a fractional uncertainty of order 10^{-16} was obtained [9].

In the twentieth century, continuous operation was possible with commercial atomic clocks. The primary atomic clocks (beam or atomic fountain) had the role to evaluate the accuracy of the time and frequency produced by these commercial atomic clocks. Now, atomic clocks with atomic fountains can be operated continuously. However, a fractional uncertainty lower than that for an atomic fountain clock has been obtained with several atomic clocks operating in the optical region (see section 3.5). Therefore, the atomic clocks operating in the microwave region have been relegated to roles requiring continuous operation rather than performing measurements of highest accuracy.

Atomic clocks of chip scale have also been developed. The microwave transition frequency between *a* and *b* states can be measured also using a laser with two frequency components. Laser light is absorbed by the atoms when its frequency corresponds to the transition frequency of either *a*–*e* or *b*–*e* (*e* is the highly excited state). However, the light is not absorbed when there are two frequency components and their frequency interval equals the *a*–*b* transition frequency; this phenomenon is called electromagnetically induced transparency or EIT (figure 3.9).

Multiple frequency components are made by modulating the laser frequency; their frequency intervals are given by the modulation frequency. The microwave transition frequency is measured by monitoring the modulation frequency at which the laser light is not absorbed. With the development of the compact laser diode having low power consumption, an atomic clock using EIT of a size much smaller than the microwave wavelength was developed; this is not possible using microwave technology [10]. Figure 3.10 shows a photo of an atomic clock, called a chip-scale atomic clock (CSAC), that is only a few millimeters in size. The measurement accuracy of the CSAC is lower than the Cs atomic clocks shown above, but is much better than quartz clocks. The low-cost CSAC with its low power consumption can

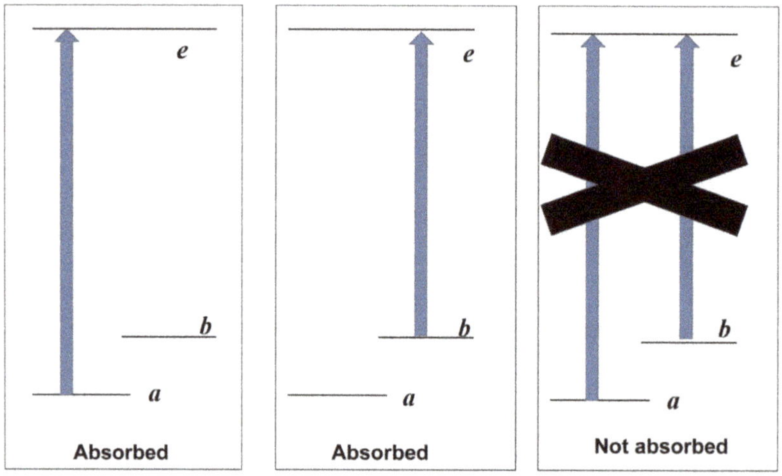

Figure 3.9. Fundamentals of EIT. Irradiating laser light that is tuned to the *a*–*e* or *b*–*e* transition is absorbed. However, the light from two laser beams tuned separately to the transitions are not absorbed.

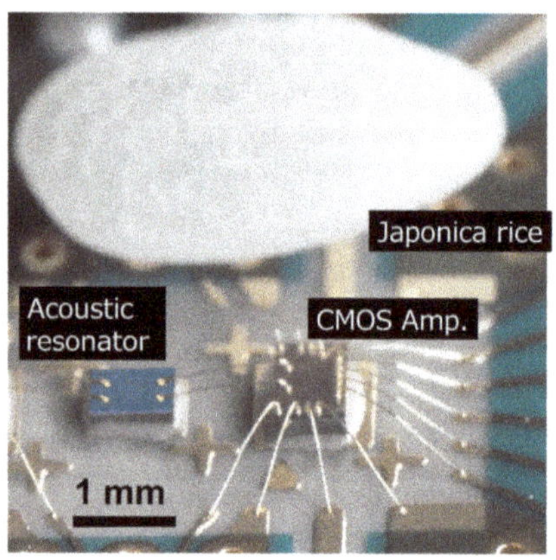

Figure 3.10. Photo of a CSAC of few millimeters in size. (Photo courtesy of M Hara, NICT, Japan.)

be inserted into devices such as mobile phones, personal computers, and GPS receivers (see section 5.1), and helps in providing greater connectivity for many people.

3.5 Optical atomic clocks

The attainable accuracy of atomic clocks based on optical transition is expected to be higher than that for microwave transitions because the time unit of 1 second can be divided into finer scales over five orders (figure 3.11).

However, optical frequencies cannot be measured using a frequency counter. To measure the optical frequency, the frequency chain system was initially developed

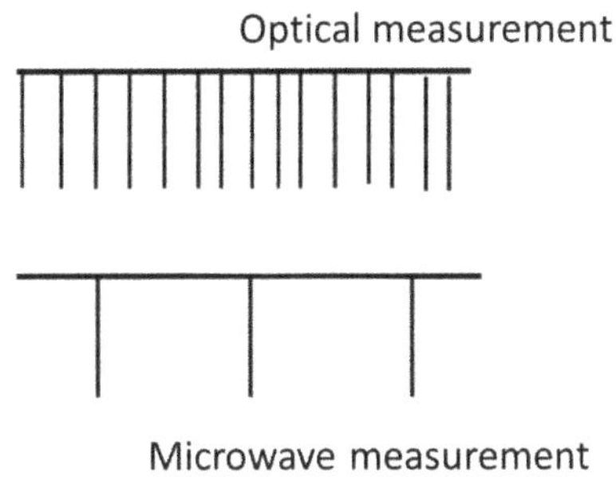

Figure 3.11. Comparison of time scales for optical and microwave measurements.

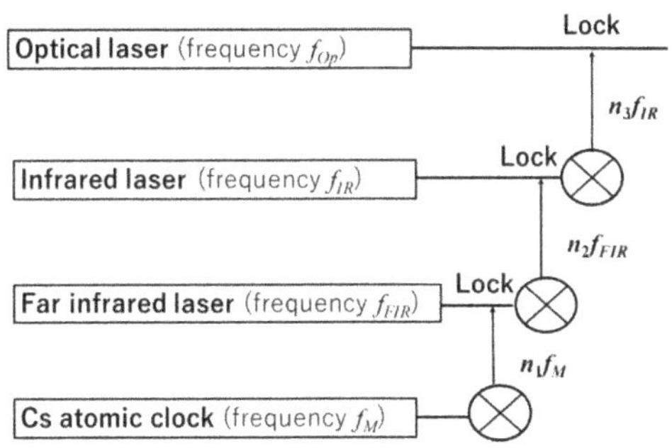

Figure 3.12. Schematic of the structure for the frequency chain constructed using a Cs atomic clock, an FIR laser, an IR laser, and an optical laser.

(figure 3.12) [11]. When an electromagnetic wave interacts with a crystal, the output includes the frequency components of integer multiples of the frequency of the incident wave. The higher-frequency source is locked to an integer multiple of the frequency of the lower-frequency source. The standard frequency given by the Cs atomic clock is multiplied and locked to the far-infrared (FIR) laser. Then the FIR frequency is also integer multiplied and locked to the infrared (IR) laser. Multiplying the IR frequency, the laser light frequency in the optical region can be locked and its frequency can be measured.

At the beginning of the twenty-first century, the frequency comb system was developed [12]. Consisting of an ultra-short pulse laser with a stable repetition frequency, its frequency components are equally spaced giving an interval of the repetition frequency (figure 3.13). The frequency components are clearly integer multiples of the repetition frequency f_{rep} and an offset frequency f_{ceo}, which can be easily measured. The laser light frequency can be measured by interferometry using the closest frequency component. In comparison with the frequency chain, the frequency comb is a much more compact apparatus and much simpler in operation. The frequency components can be stabilized by locking f_{rep} and f_{ceo} using an atomic clock in the microwave region. However, it is more common to lock one frequency component using an optical atomic clock than all frequency components being locked simultaneously.

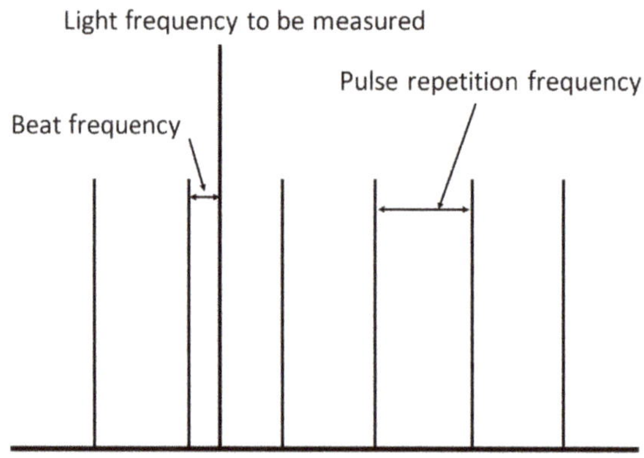

Figure 3.13. Schematic of the frequency spectrum of the frequency comb.

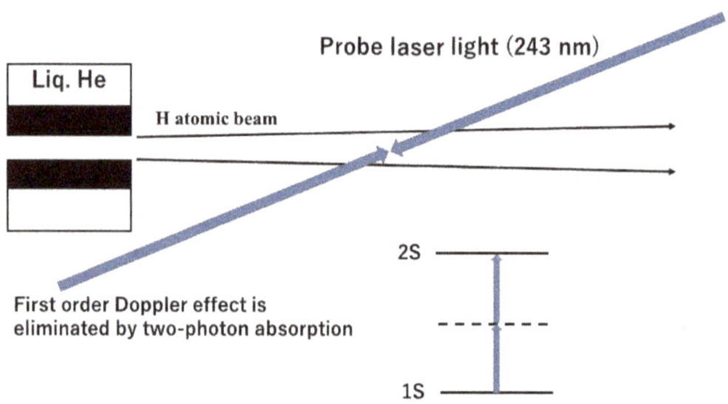

Figure 3.14. Schematic for measuring the H 1S–2 S transition frequency by two-photon absorption.

What transition in the optical region is useful for precision measurements? The high rate transitions (strong absorption) are observed with broad linewidths (see section 3.2.1) because of a high transition rate during spontaneous emissions. The linewidth of this transition (electric dipole allowed transition) is of order 1–100 MHz. Transitions with a low rate (weak absorption, electric dipole forbidden) are preferable for precision measurements. For the electric quadrupole transition, the linewidth is of several hertz. There are also transitions that are strictly forbidden in single-photon transitions but can be observed in two-photon transitions. In the latter instance, the linewidth can be narrower than 1 Hz. However, high-intensity laser light is required to induce the transition and a significant Stark shift is induced.

The Garching group measured the 1S–2S transition frequency of the hydrogen (H) atom using the frequency chain system (although they are currently using a frequency comb). The measurement was performed using a H atomic beam emitted from a nozzle cooled by liquid He, so that the velocity of the atomic beam is reduced (figure 3.14). The transition is observed by two-photon absorption induced by two laser beams (wavelength 243 nm) incident from opposing directions so that the first-order Doppler effect is eliminated. The angle between the laser light and the atomic beam was set as small as possible to ensure a long atom–light interaction time

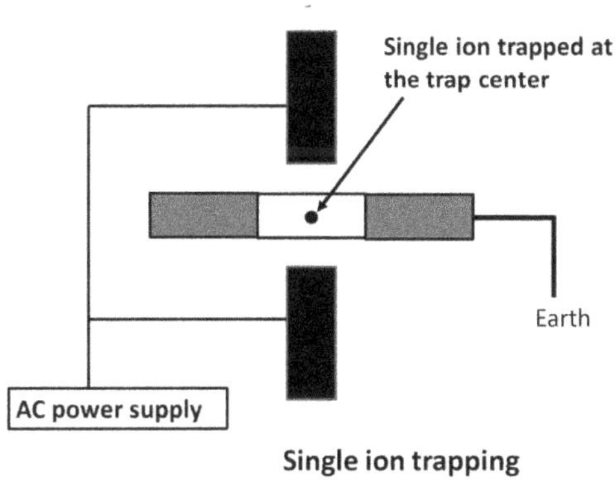

Figure 3.15. Schematic of the structure of an ion trap apparatus.

(narrow linewidth). The transition to the 2 S state is monitored by fluorescence, which is observed when the mixture between 2 S and 2 P states is induced by a DC electric field. In 1993, a fractional uncertainty of 3.7×10^{-8} was reported [13]. In 2000, this uncertainty in measurement was reduced to 1.8×10^{-14} using a Cs clock with an atomic fountain as reference [14]. The attainable accuracy is limited by the second-order Doppler shift and the Stark shift induced by the probe laser to induce the two-photon absorption. To date, lower measurement uncertainties have been obtained with some other transition frequencies (see below). However, the energy structure of the H atom is calculable and precise measurements of the H atomic transition frequencies are useful in advancing fundamental physics.

To observe the one-photon transition without the first-order Doppler shift, the atoms are required to be localized in an area smaller than the wavelength of the probe electromagnetic wave. This condition is easily satisfied with the microwave region (for the Cs atomic clock, the expansion of the atomic beam is much less than the microwave wavelength), but it is difficult with the optical region. To localize atoms in an area smaller than the optical wavelength, a strong trapping force is required that can induce the shift in the transition frequency.

Two methods were developed for the precise measurement of transition frequencies in the optical region. One is trapping an ion in an apparatus (figure 3.15) using an RF-electric field, the intensity of which is proportional to the distance from the trap center. A single ion has a vibrational motion around the trap center, where the electric field is zero. The amplitude of the vibrational motion is reduced to less than the optical wavelength by laser cooling (see appendix A). In the laser cooling stage, fluorescence is observed also from a single ion. After cooling, the cooling laser is turned off and the probe laser is switched on. To monitor the transitions induced by the probe beam, the cooling lasers are turned on again (figure 3.16). When the transition to a metastable state (an excited state with a long lifetime) is induced by the probe beam, fluorescence is not observed when the cooling laser beams are also irradiating. This measurement procedure is applicable for all ions for which laser cooling and fluorescence observations are both possible. For ions for which laser cooling is difficult, the kinetic energy is reduced through sympathetic cooling with co-trapped ions that

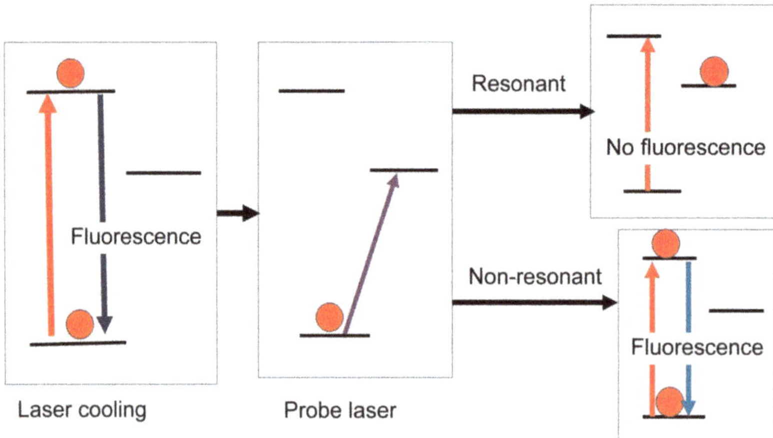

Figure 3.16. Processes in the transition of a single ion.

can be laser cooled. In addition, the energy state of the ion after irradiation by the probe beam can be monitored using the co-trapped ion [15].

Initially, the measurements were performed with the S–D transitions (orbital angular momentum 0 → 2) of alkali-like ions such as $^{40}Ca^+$, $^{88}Sr^+$, and $^{171}Yb^+$ ions. This is because laser light for laser cooling and probing can be produced using laser diodes, which are easy to obtain. This transition is split into several sub-components by magnetic field and electric field gradients. By averaging the transition frequencies for all sub-components, the Zeeman and electric quadrupole shifts are eliminated. The Stark shift induced by blackbody radiation is rather significant for these transition frequencies (shifts of order 10^{-16} at room temperature). The measurement of $^1S_0$–$^3P_0$ transition frequencies of alkali earth-like ions ($^{27}Al^+$, $^{119}In^+$) were performed subsequently. An electric quadrupole shift does not exist for this transition and the Zeeman shift at each sub-component is three orders smaller than that for the alkali-like ions. The Stark shift induced by blackbody radiation is 1–2 orders smaller than that for alkali-like ions. However, there were technical difficulties in the measurement with alkali earth-like ions because laser light in the ultra-violet region (wavelength 230–270 nm) is required.

A fractional uncertainty of the order of 8×10^{-18} was attained with the measurement of the $^{27}Al^+$ $^1S_0$–$^3P_0$ transition frequency at the National Institute of Standard and Technology (NIST, US) [15]. The accuracy is limited mainly by the second-order Doppler shift. Its fractional uncertainty can be reduced by decreasing the kinetic energy in all directions using sideband laser cooling (see appendix A). A fractional uncertainty of 3.2×10^{-18} was obtained with the ytterbium ion $^{171}Yb^+$ using the $^2S_{1/2}$–$^2F_{7/2}$ transition at the Physikalische Technische Bundesanstalt (PTB, Germany) [16]. With the same apparatus, the $^2S_{1/2}$–$^2D_{3/2}$ transition frequency of $^{171}Yb^+$ was also measured to a fractional uncertainty of 1×10^{-16}. The electric quadruple shift and the Stark shift induced by blackbody radiation in the S–F transition are smaller by a factor of 1/50 than those for the S–D transition. The frequency uncertainty in the S–F transition was estimated from the experimental

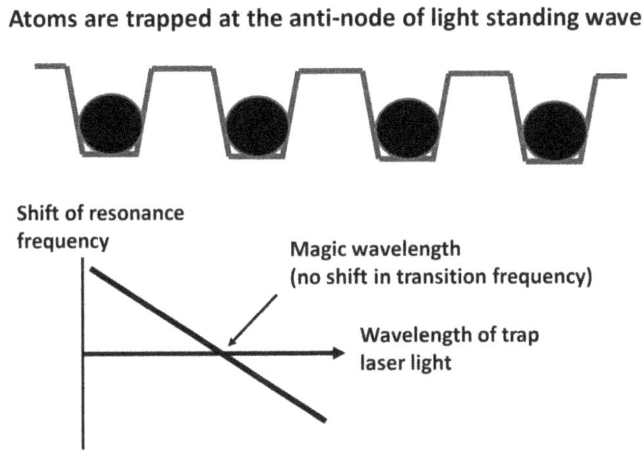

Figure 3.17. Concept behind the atomic-lattice clock. Atoms are trapped at the anti-nodes of a standing wave of a laser light with the frequency set where the transition frequency shift is zero.

uncertainty in the S–D transition frequency. Measurements were also performed with $^{199}Hg^{+}$ $^{2}S_{1/2}$–$^{2}D_{5/2}$ [17], $^{88}Sr^{+}$ $^{2}S_{1/2}$–$^{2}D_{5/2}$ [18] with fractional uncertainties of order 10^{-17}. The frequency of the $^{119}In^{+}$ $^{1}S_{0}$–$^{3}P_{0}$ transition has the theoretical potential to be measured with a fractional uncertainty of order 10^{-18}, although the obtained fractional uncertainty is currently of order 10^{-15} [19].

A method was also developed to measure the transition frequencies of neutral atoms (figure 3.17). The atoms are laser cooled (see appendix A) and trapped at the anti-node of the standing wave of the laser light (an optical lattice). Measurements are performed using many atoms to achieve low statistical uncertainty with a short measurement time. The trap laser light induces a Stark shift in the transition frequency, which can be positive or negative depending on the wavelength of the trap laser. Choosing the trap laser wavelength—called the magic wavelength—where the second-order Stark shift at the upper and lower states are equal, the shift in the transition frequency is then eliminated. When the trap laser wavelength is shifted from the magic wavelength, it is less than 10^{-18} with a shift of 1 MHz. A higher-order Stark shift exists also with the magic wavelength but is less than 10^{-18} when the atoms are trapped by a potential of 10 μK. There is a Zeeman shift of order 1 kHz G^{-1}, which is eliminated by averaging the transition frequencies of different sub-components. The accuracy is mainly limited by the Stark shift induced by blackbody radiation (order of 10^{-15} at room temperature). After the probe beam has irradiated the atoms for a period, the population in the upper state is monitored by fluorescence induced by the laser light, which is tuned to the transition from this state. We call this type of atomic clock an 'atomic-lattice clock'.

An accuracy of order 10^{-18} has been obtained for the ^{87}Sr $^{1}S_{0}$–$^{3}P_{0}$ transition frequency at RIKEN-University of Tokyo (Japan) and Joint Institute for Laboratory Astrophysics (JILA, USA) [20, 21]. To reduce the uncertainty induced by blackbody radiation, the group at RIKEN-University of Tokyo used a cryogenic chamber (liquid N_2 temperature); the group at JILA measured the temperature

Table 3.1. Transition frequencies and their uncertainties recommended at the CCTF 2015 meeting.

Transition	Frequency (Hz)	Uncertainty
$^{27}Al^{+}$ $^{1}S_0$–$^{3}P_0$	1 121 015 393 207 857.3	1.9×10^{-15}
$^{199}Hg^{+}$ $^{2}S_{1/2}(F = 0)$–$^{2}D_{5/2}(F = 2)$	1 064 721 609 899 145.30	1.9×10^{-15}
$^{171}Yb^{+}$ $^{2}S_{1/2}(F = 0)$–$^{2}F_{7/2}(F = 3)$	642 121 496 772 645.0	6.0×10^{-16}
$^{2}S_{1/2}(F = 0)$–$^{2}D_{3/2}(F = 2)$	688 358 979 309 308.3	6.0×10^{-16}
$^{88}Sr^{+}$ $^{2}S_{1/2}(F = 0)$–$^{2}D_{5/2}(F = 2)$	444 779 044 095 486.6	1.6×10^{-15}
$^{40}Ca^{+}$ $^{2}S_{1/2}(F = 0)$–$^{2}D_{5/2}(F = 2)$	411 042 129 776 398.4	1.2×10^{-14}
$^{119}In^{+}$ ^{1}S–$^{3}P_0$	1 267 402 452 899 920	3.6×10^{-13}
H 1S–2S (two-photon)	1 233 030 706 593 514	9×10^{-15}
^{87}Sr $^{1}S_0$–$^{3}P_0$	429 228 004 229 873.2	5×10^{-16}
^{171}Yb $^{1}S_0$–$^{3}P_0$	518 295 836 590 864.0	2×10^{-15}
^{199}Hg $^{1}S_0$–$^{3}P_0$	1 128 575 290 808 154.8	6×10^{-16}

distribution to very high accuracy. The measurement of the ^{87}Sr $^{1}S_0$–$^{3}P_0$ transition frequency has been performed at many other institutes in France, Germany, the US, Japan, the UK, and Italy. The idea was applied also to the $^{1}S_0$–$^{3}P_0$ transition frequencies of ^{171}Yb and ^{199}Hg atoms. The ^{199}Hg transition is useful because the Stark shift induced by blackbody radiation is one order smaller than that for the ^{87}Sr transition. However, it took time to achieve as the trap and probe lasers must operate in the ultra-violet region. The ratio of the ^{87}Sr and ^{171}Yb transition frequencies is given with a fractional uncertainty of 5×10^{-17} [22].

Measurements with a single trapped ion provide advantages in lowering the systematic uncertainty because the Stark shift induced by blackbody radiation is smaller than the transition frequencies of the atomic-lattice clocks. Lattice clock measurements with many atoms similarly provide more advantages in lowering the statistical uncertainty with a short measurement time than measurements with a single trapped ion.

An international meeting called the Consultative Committee for Time and Frequency (CCTF), held at the Bureau International des Poids et Mesures (BIPM) every three years, discusses recommendations of transition frequencies. Table 3.1 lists the values of the absolute transition frequencies and their uncertainties recommended in 2015. The uncertainties given by the CCTF are much larger than the estimations given by each experimental group because the recommended transition frequencies are absolute frequencies based on the current frequency standard given by the ^{133}Cs hyperfine transition frequency (at least until 2026), that have much larger uncertainty than that of the atomic transition frequencies in the optical region. Direct comparison with the Cs clock with an atomic fountain seems the best method to obtain the absolute frequency, the minimum fractional uncertainty being 5×10^{-16}. The frequency difference measured by the direct comparison between different atomic clocks in the optical region (see section 3.8) is often much less than the difference in absolute frequencies.

3.6 Precision measurement of molecular transition frequencies

Precision measurements of transition frequencies have been described for the atomic electric transitions (neutral atoms and ions). In this section, we consider the possibility for precision measurements of molecular transition frequencies. The energy structure of molecules is much more complicated than atoms, having vibrational and rotational energy structures (figure 3.18). Precision measurements of molecular transition frequencies are more difficult than those of atomic transition frequencies because (i) localizing molecules in a selected state and (ii) reducing the kinetic energy by laser cooling are both difficult to perform (see appendix A).

These difficulties can be overcome with molecular ions co-trapped with atomic ions in a linear trap (a trap formed by a RF-electric field given by linear electrodes in the perpendicular directions and by a DC electric field in the parallel direction). By laser cooling the atomic ion, the kinetic energy of the molecular ions is also reduced. As depicted in figure 3.19, cold ions form a string crystal along the axis, where the trap electric field is zero. Recently, a quantum-logic method was developed to localize the molecular ions in a selected state [23].

The energy shifts induced by the electric and magnetic fields (the Stark and Zeeman shifts) are quite significant for most molecular states. These energy shifts have significant dependence on the angular momentum (electron spin, nuclear spin, molecular rotation) of the states. For the vibrational transition frequencies keeping

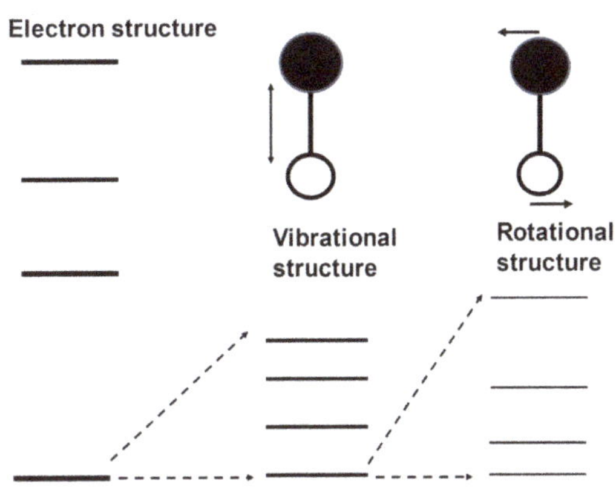

Figure 3.18. Schematic of the energy structure of diatomic molecules with vibrational rotational states.

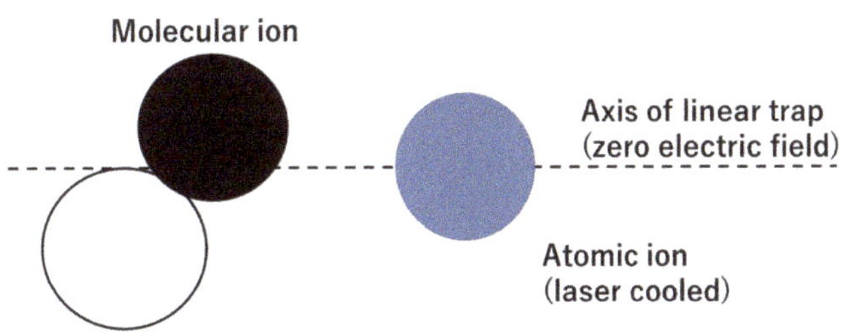

Figure 3.19. Schematic of a string crystal of one molecular ion and one atomic ion.

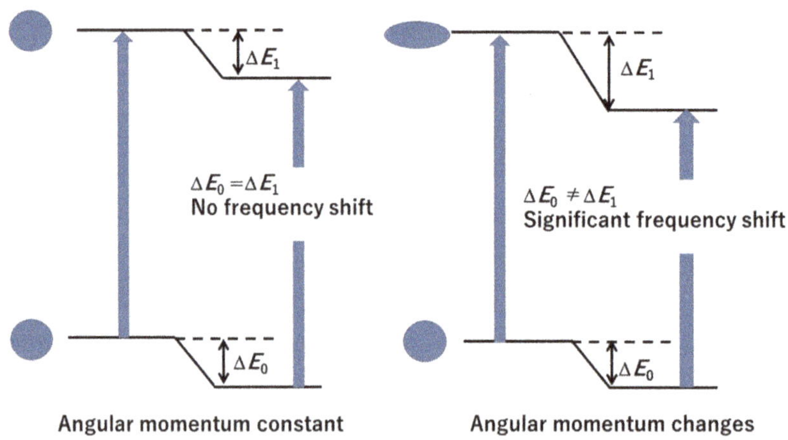

Figure 3.20. Frequency shifts induced by an electric field or a magnetic field for the upper and lower states in molecular transitions. With the angular momentum quantum numbers remaining constant, the two frequency shifts cancel each other out.

the angular momenta constant (pure vibration), the energy shifts for the upper and lower states cancel by more than 99 % (figure 3.20). We perceive the molecular shape as being unchanged during pure vibrational transitions, although the vibrational state changes the interatomic distances by a few percent. For example, molecules in the lowest rotational state are spherical for any vibrational state. Measurements with fractional uncertainties lower than 10^{-16} seem possible with pure vibrational transition frequencies.

Several proposals have been given for the precise measurement of pure vibrational transition frequencies of diatomic molecular ions [24–26]. The pure vibrational transitions of homonuclear diatomic molecular ions (N_2^+, O_2^+) offer more advantages than those of heteronuclear molecular ions because the Stark shift is smaller. Note also that blackbody radiation can induce unhelpful transitions in heteronuclear diatomic (polar) molecular ions, which is not possible for homonuclear molecular ions. For the pure vibrational transition between the lowest rotational states of homonuclear diatomic molecular ions, the fractional uncertainty of order 10^{-18} seems to be attainable. However, these transitions are one-photon forbidden and therefore two-photon transitions must be used requiring then a high-intensity probe laser to excite; hence, a significant Stark shift is induced. Using a method called hyper Ramsey spectroscopy, this Stark shift induced by the probe laser is suppressed [27].

Precision measurements are possible also with the dipole-allowed vibrational transition, which changes the rotational state by ± 1. The Stark, Zeeman, and electric quadrupole shifts are eliminated by averaging the transition frequencies for all subcomponents.

Measuring the pure vibrational transition frequencies of $^{174}Yb^6Li$, $^{88}Sr^6Li$, and $^{40}Ca^6Li$ molecules is also possible [28, 29]. Molecules are produced bonded to ultracold atoms and trapped in an optical lattice with the magic wavelength. However, producing such molecules with ultra-low kinetic energies has never been

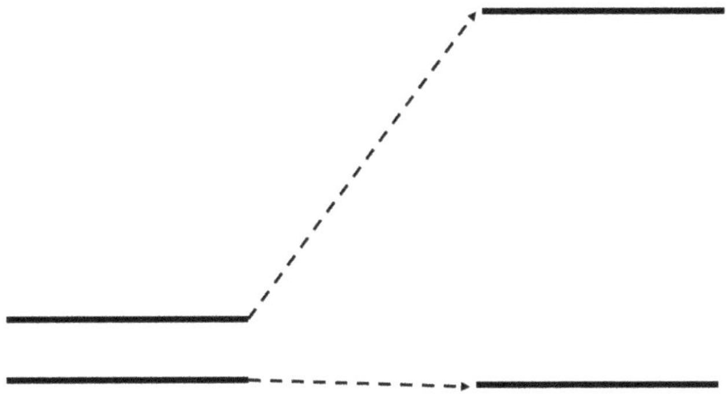

Figure 3.21. Transition frequencies of highly charged ions are 2–3 orders higher than the same transition frequencies of neutral atoms.

attained. Therefore, measurements with molecular ions appear to be more realistic.

Precise measurements of molecular transition frequencies are useful to observe phenomena which cannot be observed with atomic transitions, and can be helpful for the development of physics beyond the accepted Standard Model of particle physics.

3.7 Other possible clocks

The transition frequencies of highly charged ions have recently been considered to be useful for precise measurement. With highly charged ions, the interaction between a nucleus and electron is much stronger than that for atoms with the same number of electrons. Therefore, the transition frequencies of highly charged ions are 2–3 orders higher than the same transition frequencies of neutral atoms with the same electron number (figure 3.21).

The Stark shift induced by, for example, the electric field produced by the probe laser and blackbody radiation, is much less than that for neutral atoms because energy gaps between the different electron states are much larger than those for neutral atoms. Moreover, the second-order Zeeman shift is less than 10^{-11} G^{-2}; therefore, the Zeeman shift is negligibly small for sub-components where the first-order Zeeman shift is zero ($M = 0$).

There is a technical difficulty in controlling highly charged ions because the initial kinetic energy is of order 1 MK. Laser cooling of a highly charged ion is not realistic because the electron transition cannot be induced by laser light operating in the optical region. However, Ar^{13+} ions were cooled in interactions with laser cooled Be^{+} ions and trapped in a crystal [30].

Precise measurements of nuclear transition frequencies are also of interest. The energy gap between different nuclear states is much larger than that for atomic states because of the strong repulsive force between protons bounded in an area of femtometer size (figure 3.22). Most of the nuclear transition frequencies are in the

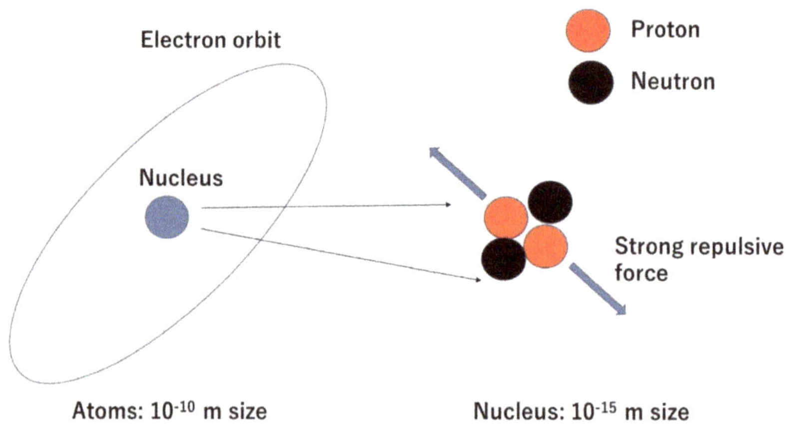

Figure 3.22. Schematic of the nuclear structure. The strong repulsive forces between protons are bounded in an area five orders smaller than atoms.

γ-ray region, where frequency measurements are difficult. Nevertheless, the thorium-299 nucleus (^{299}Th) has a transition in the vacuum ultra-violet (VUV) region (1.89 PHz, 159 nm). Such frequencies can be measured using a frequency comb [31]. With a suitably chosen electron state, this nuclear transition is free from the first-order Zeeman shift and the second-order Stark shift. Therefore, the systematic uncertainty is much lower than that for atomic transition frequencies.

Limited by the spontaneous emission transition, the linewidth is less than 0.1 MHz. At the PTB, Germany, a project is under way to measure the nuclear transition frequency of ^{299}Th using trapped and laser cooled Th^{3+} ions. A fractional uncertainty lower than 10^{-20} may be attained. However, experimental difficulties remain in that the transition frequency is known only to 10%. There are also technical problems to obtain and transmit laser light in the VUV region.

Nuclear transition frequencies can be measured also by using Th atoms doped in a crystal. However, the shift associated with the crystal field is proportional to the temperature of the crystal and may be much larger than 10^{-15} K^{-1}.

3.8 Comparison between atomic clocks

How can we say that our atomic clocks are accurate? To have reliability in accuracy, a comparison with other atomic clocks is required. It is not difficult to compare different atomic clocks in the same laboratory. However, reliability of this sort is not enough because atomic clocks developed in the same laboratory may have common errors. Comparisons between atomic clocks in different laboratories is required to establish a true reliability.

A direct common view of a satellite (see figure 3.23) is useful in comparing atomic clocks at distant locations. Laboratories A and B can compare the time of their atomic clocks T(A, B) with a time given by an atomic clock in a satellite T(S) and obtain $\Delta T(A, B) = T(A, B) - T(S)$. The time difference between atomic clocks in laboratories A and B is given by $\Delta T(A) - \Delta T(B)$. However, taking the propagation

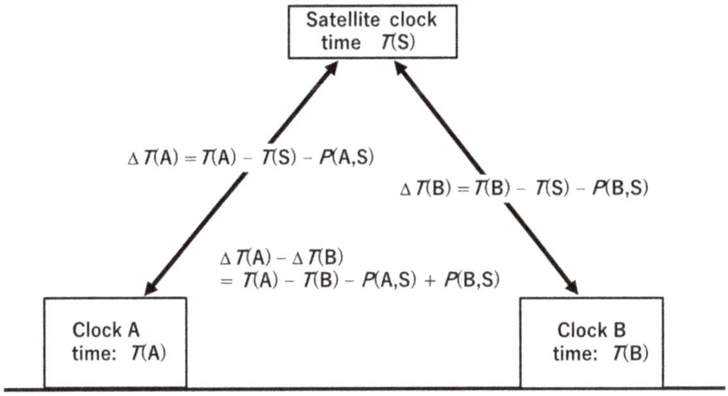

Figure 3.23. Comparison between two clocks A and B via a satellite in direct common view. Both clocks are compared with a clock in a satellite that has direct views of both places. There is an uncertainty error in the propagation time.

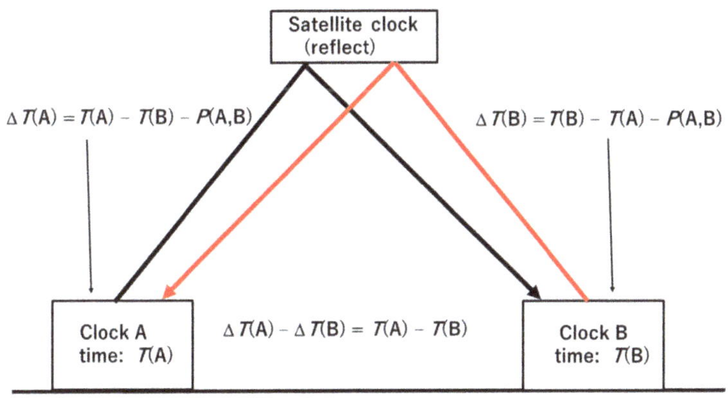

Figure 3.24. Two-way satellite time and frequency transfer method. With this method, the error from the propagation time is eliminated.

time of the radio signals between clocks A, B and the clock in the satellite, $P(\mathrm{A, S})$ and $P(\mathrm{B, S})$, there is an error of $P(\mathrm{A, S}) - P(\mathrm{B, S})$.

To eliminate the error from the propagation time, a two-way satellite time and frequency transfer method is useful. A radio signal from clock B is reflected by the satellite and received by clock A (figure 3.24). The time difference observed at clocks A and B are

$$\Delta T(\mathrm{A}) = T(\mathrm{A}) - T(\mathrm{B}) - P(\mathrm{A, B}) \qquad \Delta T(\mathrm{B}) = T(\mathrm{B}) - T(\mathrm{A}) - P(\mathrm{A, B}) \tag{3.14}$$

and

$$\Delta T(\mathrm{A}) - \Delta T(\mathrm{B}) = T(\mathrm{A}) - T(\mathrm{B}). \tag{3.15}$$

With this method, the error from the propagation time is eliminated. The frequency difference between the ^{87}Sr lattice clocks in Germany and Japan was measured to be

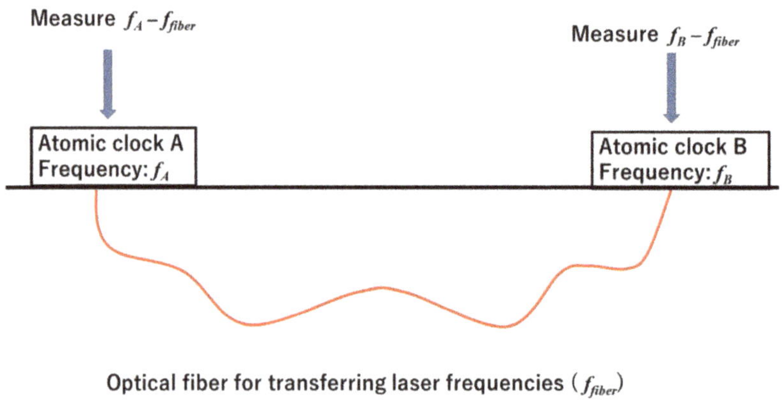

Figure 3.25. Comparison between two atomic clocks A and B by optical fiber link.

$(1.1 \pm 1.6) \times 10^{-15}$ [32]. The frequency ratio between the ^{87}Sr lattice clock in Japan and ^{171}Yb lattice clock in South Korea was measured yielding a fractional uncertainty of 5.8×10^{-16} [33].

To compare the atomic clocks in the optical region, an optical fiber connecting the different laboratories is useful (figure 3.25). Each laboratory compares their probe laser frequencies with the frequency of the laser light, which is transmitted to other places via the fiber connection. The power loss of laser light in propagating along the optical fiber is low at the wavelength of 1.4–1.6 μm.

The ^{87}Sr lattice clocks in Germany and France (1400 km apart) were compared yielding a fractional uncertainty of 5×10^{-17} [34]. With this comparison, we can observe the influence of the difference in gravitational potential, including Earth's tides. Currently, ^{87}Sr lattice clocks in Germany, France, and the UK are linked by optical fiber. It is possible to test the theory of special relativity by observing the time dilation given by the relative velocities between the different places (recall that, at different latitudes, Earth's rotation produces different velocities) [35]. In the future, a comparison is to be performed between atomic clocks in different continents.

An optical fiber link is useful in sending laser light of a standard frequency to many other laboratories, thereby making rapid progress in many fields in regard to fundamental physics (see chapter 4).

Exercises

3.1 The statistical uncertainty of transition frequency is given by

$$\frac{\Delta f}{f}\sqrt{\frac{\tau}{NT}},$$

where Δf is the spectrum linewidth, f the transition frequency, τ the one-cycle measurement time, N the number of measured atoms, and T the measurement time. Estimate the measurement time required to get the statistical uncertainty lower than 10^{-16} for a Cs atomic clock with a fountain and for a Sr atomic clock with optical lattice. We find that the transition frequencies in the optical region afford many more advantages than the microwave region;

Cs atomic clock with fountain	$\Delta f = 1\text{Hz}, f = 9.2 \times 10^9\text{Hz}, \tau = 1s, N = 10^5$
Sr atomic clock with lattice	$\Delta f = 1\text{Hz}, f = 4.3 \times 10^{14}\text{Hz}, \tau = 1s, N = 10^4$.

3.2 Estimate the second-order Doppler shift $\Delta f/f$ with the hydrogen maser. The average velocity of the hydrogen atom at room temperature is $v = 2700$ m s^{-1}. Light velocity is $c = 3 \times 10^8$ m s^{-1}.

Solutions

3.1 Cs atomic clock 1.2×10^7 s = 136 days; Sr atomic clock 0.05 s.
3.2 $(\Delta f/f) = -(v/c)^2/2 = -4 \times 10^{-11}$.

References

[1] Vanier J and Audoin C 1989 *The Quantum Physics Atomic Frequency Standards* (Bristol: Adam Hilger), p 610
[2] Vanier J and Tomescu C 2017 *The Quantum Physics Atomic Frequency Standards: Recent Developments* (Boca Raton, FL: CRC Press), p 1
[3] Vanier J and Tomescu C 2017 *The Quantum Physics Atomic Frequency Standards: Recent Developments* (Boca Raton, FL: CRC Press), p 15
[4] Hasegawa A *et al* 2004 *Metrologia* **41** 257
[5] Clairon A, Salomon C, Guellati S and Phillips W 1991 *Europhys. Lett.* **16** 165
[6] Heavner T P *et al* 2005 *Metrologia* **42** 411
[7] Wikipedia 2018 Hydrogen maser
[8] Camparo J 2007 *Phys. Today* **60** 33
[9] Ovchinnikov Y and Marra G 2011 *Metrologia* **48** 87
[10] Knappe S *et al* 2005 *Opt. Lett.* **30** 2351
[11] Evenson K M *et al* 1973 *Appl. Phys. Lett.* **22** 192
[12] Adler F *et al* 2004 *Opt. Express* **12** 5872
[13] Schmidt-Kaler F *et al* 1993 *Phys. Rev. Lett.* **70** 2261
[14] Niering M *et al* 2000 *Phys. Rev. Lett.* **84** 5496
[15] Chou C W *et al* 2010 *Phys. Rev. Lett.* **104** 070802
[16] Huntemann N *et al* 2016 *Phys. Rev. Lett.* **116** 063001
[17] Tanaka U *et al* 2003 *J. Phys. B: At. Mol. Opt.* **36** 545
[18] Dube P *et al* 2017 *Metrologia* **54** 290
[19] Ohtsubo N 2017 *Opt. Express* **25** 11725
[20] Ushijima I *et al* 2015 *Nat. Photon.* **9** 185
[21] Nicholson T L *et al* 2015 *Nat. Commun.* **6** 6896
[22] Nemitz N *et al* 2016 *Nat. Photon.* **10** 258
[23] Leibfried D 2012 *New J. Phys.* **14** 023029
[24] Kajita M and Moriwaki Y 2009 *J. Phys. B: At. Mol. Opt.* **42** 154022
[25] Kajita M *et al* 2014 *Phys. Rev.* A **89** 032509
[26] Kajita M 2017 *Phys. Rev.* A **95** 023418
[27] Yudin V I *et al* 2010 *Phys. Rev.* A **82** 011804(R)

[28] Kajita M *et al* 2011 *Phys. Rev.* A **84** 022507
[29] Kajita M *et al* 2013 *J. Phys. B: At. Mol. Opt.* **46** 025001
[30] Lopez-Urrutia J R C 2016 *J. Phys.: Conf. Ser.* **723** 012052
[31] Peik E and Tamm C 2003 *Europhys. Lett.* **61** 181
[32] Hachisu H *et al* 2014 *Opt. Lett.* **39** 4072
[33] Fujieda M *et al* 2018 *IEEE Trans. Ultrason. Ferroelectr. Freq. Control.* **65** 973–8
[34] Lisdat C *et al* 2016 *Nat. Commun.* **7** 12443
[35] Gersl J *et al* 2015 *Metrologia* **52** 552

Chapter 4

The role of precision measurements in future developments in physics

4.1 Introduction

Precision measurements of atomic and molecular transition frequencies are expected to contribute to the development of physics in various ways including the following (figure 4.1):

1. Precise measurements of transition frequencies give the most accurate standard for time keeping.
2. With the speed of light a universal constant, a precision measurement of the frequency of light gives a precise wavelength, and therefore an accurate length standard.
3. Precise measurements of transition frequencies of atoms and molecules give detailed information about their energy structure, as well as the characteristics associated with their configuration of elementary particles.
4. Precise measurements of time and frequency are important to astronomical observations such as determining the position and velocities of stars.

This chapter introduces several examples of current physics research in which precision measurements of transition frequencies have provided important contributions.

4.2 Confirmation of theory of relativity

4.2.1 Second-order Doppler shift and gravity redshift

Recalling section 2.4, time goes slower for an object that is either traveling with velocity (second-order Doppler shift) or in a higher gravitational potential (gravity redshift). As demonstrated in section 3.2, these effects contribute to limiting the accuracy of atomic clocks.

doi:10.1088/978-0-7503-2124-2ch4

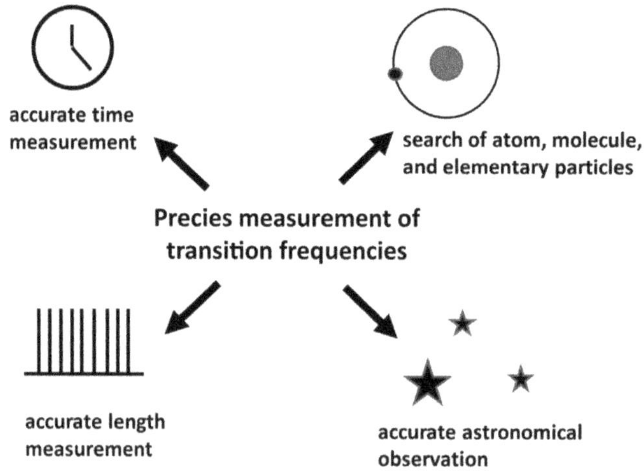

Figure 4.1. Role of precision measurements of transition frequencies.

Nevertheless, atomic clocks are useful tools in confirming relativistic effects. With Cs atomic clocks using atomic beams, second-order Doppler shifts of order 10^{-12} can be observed. With an atomic beam selected for a magnetic state, only atoms with a limited range in velocity contribute to the measurement [1]. This range in atomic velocity depends on the structure of the magnet selector, and there is a difference between measurements using atomic clocks with different velocity ranges. Making a correction using the second-order Doppler shift from a measured velocity distribution reduces the frequency difference significantly. The second-order Doppler shift is observed clearly from a comparison between Cs atomic clocks with an atomic beam and an atomic fountain (second-order Doppler shift $< 10^{-17}$). The second-order Doppler shift induced by velocities in the range 0–40 m s^{-1} was measured using the $^{27}Al^{+}$ transition frequency, the fractional uncertainty being of order 10^{-18} [2].

The gravitational redshift is observed as a change in the transition frequency with changing altitude (10^{-16} m^{-1}). It was measured by comparing the frequencies of a Cs atomic clock when located in a valley and on a nearby mountain [3]. The change in the gravitational redshift with the difference of the altitude of 15 cm was observed using the $^{27}Al^{+}$ transition frequency [2].

4.2.2 Detection of the gravitational wave

In section 2.4, gravity was described as a distortion of space and propagates as a wave with the speed of light. For gravitational waves (with spin polarization of 2), the size in the direction perpendicular to the propagation repeatedly expands and contracts (figure 4.2). Its effect has been very difficult to observe, because the change in size is so small being 10^{-21}, corresponding to observing a change in the size of an atom over the distance between the Sun and Earth.

However, a phenomenon was discovered to confirm the existence of gravitational waves indirectly. The orbital period of a binary neutron star PSR 1913 + 16 (7.75 h) decreased by 7.65×10^{-5} s yr^{-1} [4] (figure 4.3). They are expected to collapse 300

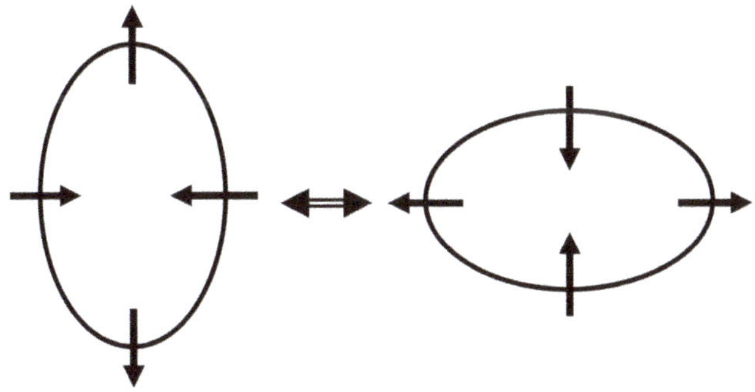

Figure 4.2. Gravitational waves undergo repeated expansion and contraction in size in the direction perpendicular and parallel to its direction of propagation direction.

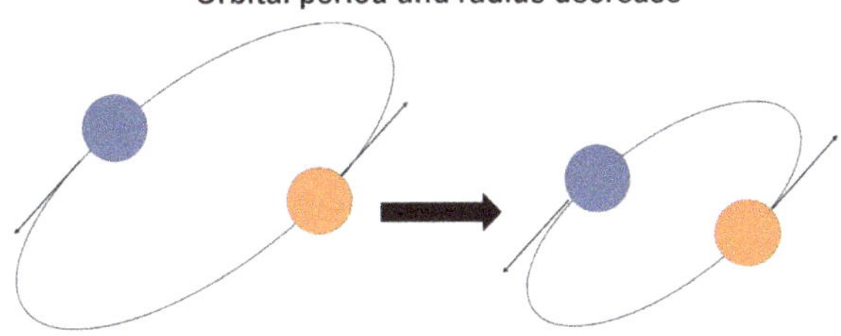

Figure 4.3. Orbital period and radius of the binary neutron star decrease because energy is lost with the generation of a gravitational wave.

million years from now. This result shows the loss of energy in orbital motion and it is reasonable to state that this energy loss is in the form of radiating gravitational waves. The measured variation in the orbital period is in good agreement with the estimation obtained from the theory of general relativity to within 0.2%. The accuracy of the atomic clock contributed to measuring the orbital period with high precision (present accuracy of 1.5×10^{-5} s).

A laser interferometer was constructed for the direct detection of gravitational waves. The apparatus (figure 4.4) consists of a laser (Nd:YAG laser, wavelength: $\lambda = 1064$ nm) that shines light onto a half mirror, which splits and redirects the light into two perpendicular directions. In the arms of the interferometer (resonators of length $L_{1,2}$), the light is reflected many times ($n_{1,2}$ times). Both beams are brought together in the half mirror again to form an interference pattern exhibiting a phase difference of

$$\phi = 2\pi(n_1L_1 - n_2L_2)/\lambda. \tag{4.1}$$

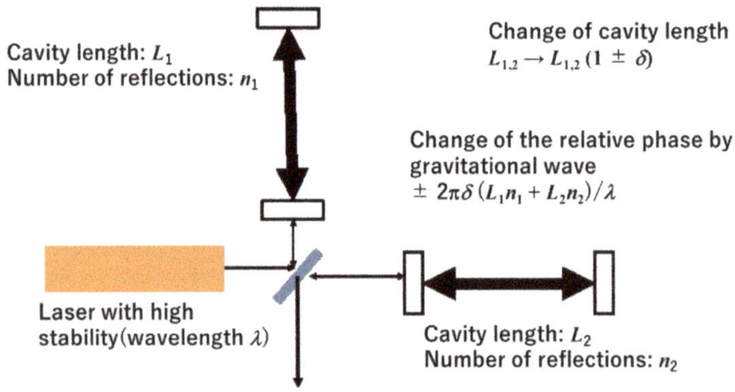

Figure 4.4. Laser interferometry system to detect gravitational waves.

When the gravitational wave changes the cavity length by $L_{1,2} \rightarrow L_{1,2}(1\pm\delta)$, there is a change in phase difference of

$$\delta\phi = 2\pi\delta(n_1L_1 + n_2L_2)/\lambda. \tag{4.2}$$

The first direct observation of the gravitational wave was made on 14 September 2015 at the Laser Interferometer Gravitational-wave Observatory (LIGO) in Hanford (Washington, US) and Livingston (Louisiana, US) [5]. Each has a sensitivity to detect the change in the length of the 4 km cavity of one-thousandth of the width of a proton (~1 am). The detected waveform (frequency 30–250 Hz) matched the prediction of general relativity for a gravitational wave emerging from the merger of a pair of blackholes of 36 and 29 solar masses. The frequency stability (10^{-15}) of the laser light was important in attaining high sensitivity to the slight changes in cavity length. The laser power also must be stabilized to 10^{-6}, because the fluctuation in radiation pressure changes the cavity length. Precision measurements in the slight difference in detection times at both laboratories are important in finding the source of the gravitational wave. The source was found to be 1.3 billion light years away.

LIGO detected three more gravitational waves from merges of blackholes. On 26 December 2015, the second gravitational wave was detected from the merger of two blackholes, each of 21 solar mass, 1.4 billion light years away. The third gravitational wave was detected on 4 January 2017 from the merger of a 49 solar-mass blackhole about 3 billion light years away. The fourth on 14 August 2017 was from the merger of blackholes of 12 and 7 solar masses 1 billion light-year away. It was also detected at the VIRGO interferometer (Cascina, Italy). Its origin has since been determined in more detail [6]. On 17 August 2017, a gravitational wave from two merging neutron stars was detected by LIGO and VIRGO, which was observed also in the electromagnetic radiation [7].

In the near future, gravitational waves will be detected at the Kamioka Gravitational Wave Detector (KAGRA, Japan) [8]. Being constructed underground near Kamioka, the influence of Earth's vibration is minimized. The cavities are constructed with sapphire mirrors cooled to 20 K to reduce thermal noise. When a gravitational wave is detected in many places, more detailed information about their origin can be obtained.

Detection of gravitational waves can give important information about the Universe, which cannot be obtained by observations with light, microwave or x-rays. In particular, details about the Universe before it became transparent to radiation (380 000 years after the Big Bang) may be obtained from gravitational waves. However, the wavelength of the gravitational wave from the Big Bang is expected to be very long and a much longer cavity than currently in LIGO and VIRGO is required. For the future, the Laser Interferometer Space Antenna (LISA), led by European Space Agency (ESA) [8], is a project that will launch three satellites to be separated by millions of kilometers. LISA's satellites are to be linked via laser beams and to act as a single interferometer.

4.2.3 Search for the gravitational effects at the micro-scale

The unification of the theory of general relativity and quantum mechanics has never been attained. Moreover, there is mystery as to why the gravitational force is extremely small in comparison with the other three interactions (strong nuclear, electromagnetic, and weak nuclear force). Superstring theory has been proposed (not confirmed yet) to resolve these questions. This theory considers particles to be a string having a finite size rather than being point-like (figure 4.5); particle mass is given by the vibrational energy of the string. With this idea, the gravitational force at micro-scales can be much stronger than that obtained from the law of universal gravitation, which has been confirmed to distances larger than 10 μm [9, 10] (figure 4.6). Alternatively, the gravitational force at micro-scales is believed to be a fifth force.

The gravitational force has been studied at distances much larger than atomic sizes, and hence the electric force from neutral atoms or molecules is negligible. In contrast, the electromagnetic force is dominant over distances of atomic size. The gravitational force at micro-scales is difficult to detect because its magnitude is much less than the electromagnetic force even if it may be much larger than that given by the law of universal gravitation.

The gravitational force needs to be scrutinized for any dependence of its potential at interatomic distances. This can be obtained from measurements of molecular binding energies in different vibrational rotational states (figure 4.7). The interatomic potential is mostly dominated by the electromagnetic interaction. However, a deviation not derivable from the electromagnetic interaction may appear with a

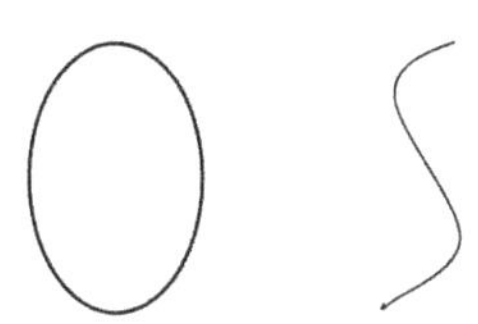

Figure 4.5. Schematic of particles as strings of finite length in the superstring model.

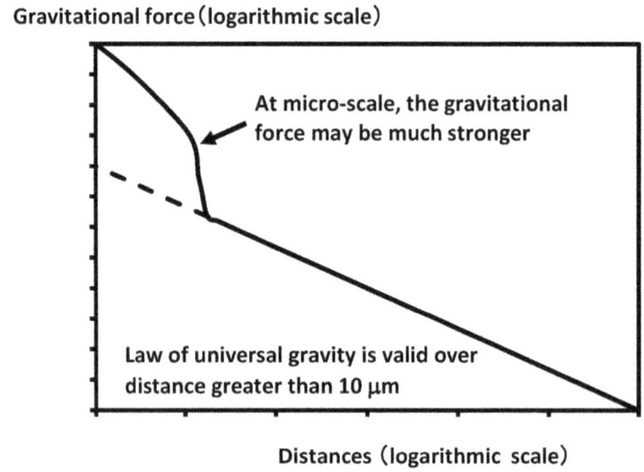

Figure 4.6. Dependence of the gravitational force on the distance between objects. The law of universal gravitation is confirmed only for distances larger than 10 μm. With smaller distances, a much stronger interaction may hold than given by Newton's law.

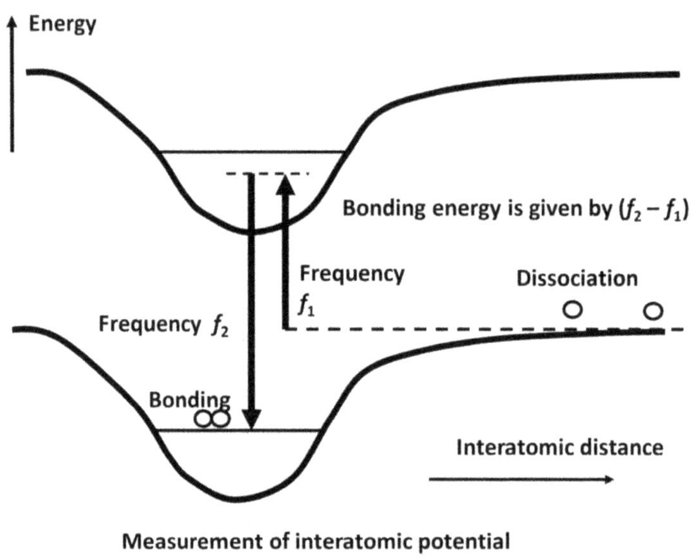

Figure 4.7. Searching for a gravitational effect with precision measurements of the interatomic potential from molecular bonding.

significant improvement in measurement accuracy. Any discrepancy would be a gravitational effect if it is proportional to the atomic mass and confirmed through measurements on different isotopes.

Although we experience only four dimensions (three spatial and one temporal), superstring theory requires the existence of ten or more dimensions. If a strong gravitational force at micro-scales (a fifth force) is found, it may also confirm the existence of extra dimensions. Then, where are the extra dimensions? Lines viewed at

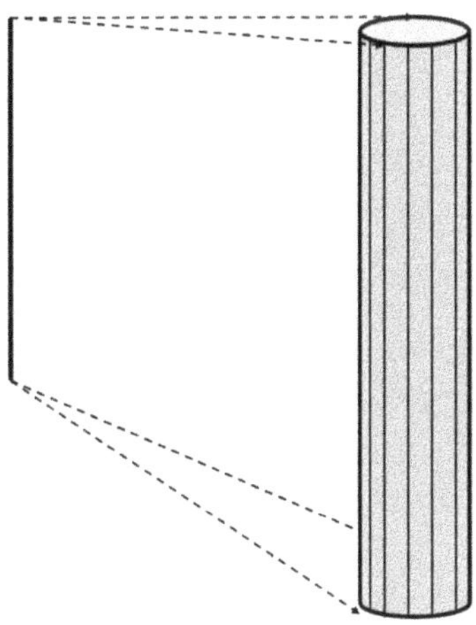

Figure 4.8. Schematic of extra dimensions. A line viewed at a distance looks one-dimensional but may be cylindrical with a finite radius when viewed close-up.

macro-scale may in reality be cylinders having an extra dimension when viewed at micro-scales (figure 4.8). The existence of extra dimensions might be confirmed from a process that does not conserve energy (and lost in hidden dimensions), and may require measurements of ultrahigh precision.

4.3 Search for symmetry violations in particles and anti-particles

4.3.1 C-, P-, and T-symmetry

Based on Newtonian mechanics, the Schrödinger equation is useful in obtaining the wavefunctions of atoms and molecules. However, the intrinsic spin of an electron is not described by this equation (cf the two-component spinor equation of Pauli, 1927). Experimentally, passing a beam of electrons through an inhomogeneous magnetic field demonstrated that the Zeeman energy shift in half of the electrons were positive and the other half were negative. In 1928, Dirac derived an equation to obtain the quantum mechanical waveform of the electron based on the theory of special relativity [11]. The Dirac equation is a 4×4 matrix equation for a wavefunction given as a four-dimensional column vector. The four solutions obtained correspond to spin $\pm 1/2$ and invariant-mass energies $\pm mc^2$.

The Dirac equation was successful in describing the electron spin but added a new mystery with the appearance of negative-mass energy. Dirac proposed a model featuring a sea of negative-mass electrons to solve this problem. With this model, the vacuum is filled with such electrons (figure 4.9); if one such electron is excited with an energy of $2mc^2$ to a positive mass state, a conventional electron is produced. The absence of a negative-mass electron is observed as a positive charge particle of the same mass as the electron. This notion of an anti-particle was confirmed by the discovery of the positron. Pair production and pair annihilation of electron and

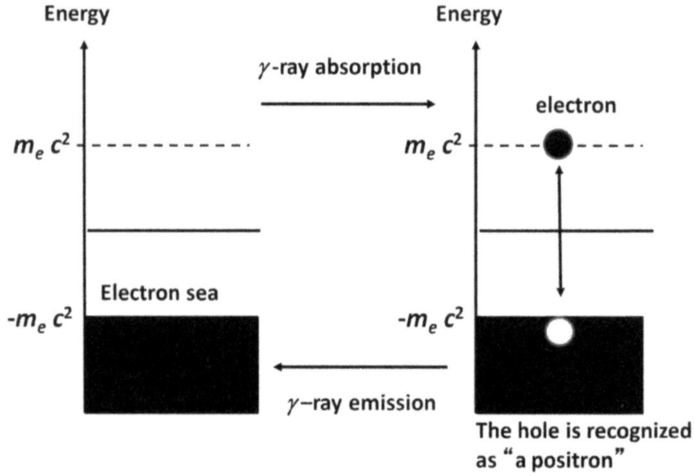

Figure 4.9. Dirac's 'electron-sea model' was later adapted to predict the existence of positrons.

positron were also observed. Anti-protons and anti-neutrons were also discovered afterwards. Currently, though, a new interpretation is considered instead of the electron-sea model. For all particles, there also exist anti-particles, for which the characteristics are identical except for a change in sign of the electric charge.

There remains an unsolved mystery. A change in the numbers of particles and anti-particles is possible only by pair production and annihilation (figure 4.10). It is not possible that only particles (anti-particles) appear or disappear. However, we see only particles in the Universe and apart from those produced in pair production no anti-particles. It is believed that both particles and anti-particles existed at the dawn of the Universe and anti-particles disappeared in pair annihilation with particles. However, it has been argued that particles dominate the Universe because their number was slightly larger than that of anti-particles (in a ratio $1+10^{-9}$:1). While possible, the mystery of the asymmetry between particle number and anti-particle number is not yet solved.

The following symmetries were considered [12].

(1) Charge symmetry (C-symmetry): Symmetry in the physical laws under charge conjugation (the interchange of particle with anti-particle).
(2) Parity symmetry (P-symmetry): Symmetry in the physical laws under mirror reflection.
(3) Time-reversal symmetry (T-symmetry): Symmetry in the physical laws under time reversal.

C-symmetry is violated because the neutrino comes only with spin $-1/2$, whereas the anti-neutrino has only spin $+1/2$. P-symmetry is also violated as experimental results from the β-decay of Co atoms found that the direction of the emitted electrons is localized in the direction anti-parallel to the nuclear spin [13]. However, CP-symmetry (symmetry associated with the simultaneous application of charge and mirror reflection) was expected to be conserved because the spin of the neutrino

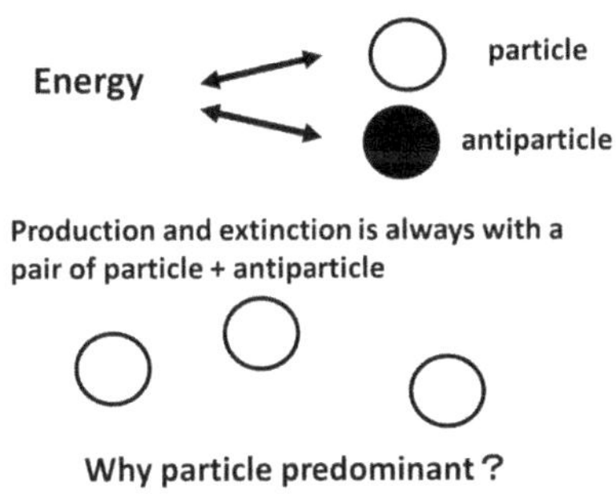

Figure 4.10. Creation and annihilation of particles and anti-particles pairs always proceeds in pairs. Nevertheless, why the Universe is dominated by just particles alone remains a mystery of modern physics.

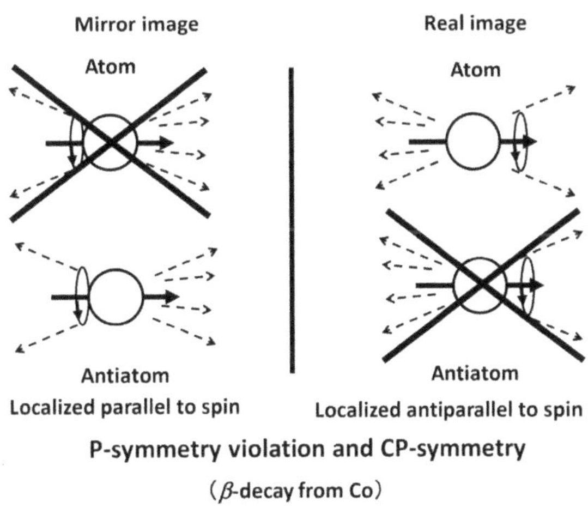

Figure 4.11. The P-symmetry violation observed with the direction of emitted electron at the β-decay of Co atom. If the inverted result were obtained with anti-atom, CP-symmetry is conserved.

under mirror reflection equals that of an anti-neutrino. The localization of the direction of an emitted positron in the β-decay of an anti-atom was expected to be in the direction parallel to the nuclear spin (figure 4.11).

In 1964, the evidence for CP-symmetry violation was demonstrated from the differences in kaon decays [14]. There are two kinds of neutral kaons K_1 and K_2: K_1 decays into two pions (CP symmetric) whereas K_2 decays into three pions (CP anti-symmetric). The decay of K_1 is 100 times faster than K_2; thus, a long beam is almost pure K_2. However, 0.2% of the kaons decay into two pions. This result shows a transformation from K_2 to K_1, and hence the CP-symmetry is violated.

More direct evidence was developed during the 1990s [15]. The neutral kaon is able to transform into its anti-particle and vice versa; however, there is a difference between the probabilities in the two directions. This result shows that not only is CP-symmetry violated but also T-symmetry; only CPT-symmetry is conserved. The CP- and T-symmetry violations are more prominent in B-meson decays (CPT-symmetry conserved) [16].

The Kobayashi–Maskawa theory shows that the violation of CP-symmetry is possible if there are six quarks [17]; the existence of all six quarks was experimentally confirmed. The violation of CP-symmetry shows that there is some difference between particles and anti-particles except for the sign of the electric charge. Nevertheless, the CP-violation determined from the Kobayashi–Maskawa theory is insufficient to account for the particle predominance in the Universe.

The CPT-symmetry is required by Lorentz invariance. This chapter introduces the research results confirming CPT-symmetry and T-symmetry violation.

4.3.2 Confirmation of CPT-symmetry

The equality of mass as well as the absolute value of the electric charge between particles and anti-particles is required for CPT-symmetry.

The equality of the S_1 = (electric charge)^4 × (mass) between proton and anti-proton was confirmed with a fractional uncertainty of 10^{-9} obtained from precision measurements of the anti-proton helium atom (figure 4.12) [18]. An anti-proton helium atom consists of a helium nucleus (charge +2), one electron (charge −1), and one anti-proton (charge −1). The orbital radius of the anti-proton is much less than that of an electron. The electron cloud, the role of which is to prevent other particles colliding with the anti-proton, does not interact with the anti-proton via the electric force. The energy level of the anti-proton is determined only by the interaction of the nucleus, which is approximately given by the formula for the electron in a hydrogen-like atom. From precision measurements of the transition frequencies, the ratio of S_1 between anti-proton and electron was obtained. After the proton-to-electron mass ratio was determined, the equality of S_1 between proton and anti-proton was confirmed within a fractional uncertainty of 10^{-9}.

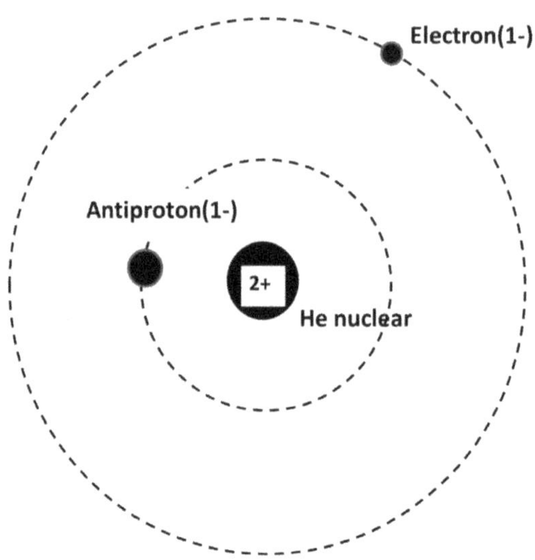

Figure 4.12. Schematic depicting the structure of an anti-proton helium atom, consisting of a He nucleus, anti-proton, and electron.

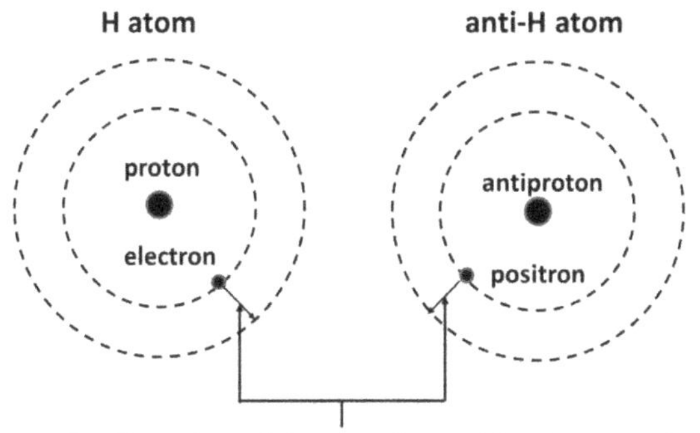

Figure 4.13. Comparison of transition frequencies of the H atom and the anti-H atom.

The equality of S_2 = |charge|/mass between proton and anti-proton was confirmed by comparing the cyclotron frequencies, set by a magnetic field of 1.95 T, and achieved an accuracy of 6.9×10^{-11} [19].

The equality of the 1S–2 S transition frequencies between the H atom and anti-H atom (consisting of an anti-proton and positron) was also confirmed (figure 4.13) with an uncertainty of 2×10^{-10} [20].

The precision measurements of the transition frequencies played an important role in confirming the CPT-symmetry. Nevertheless, there is still the possibility that CPT-symmetry violation is detected after improvements in the measurement accuracy.

4.3.3 Search for electron electric dipole moment

Recently, the search for an electric dipole moment (EDM) in the electron (eEDM) has drawn interest as direct evidence of T-symmetry violation. If the eEDM exists, there are at least two states $\pm d$ with distinct directions (figure 4.14). The electron has two spin states $s = \pm 1/2$, and there are just two states for the electron. Therefore, the possible eEDM state can be one state corresponding to the spin state: the possible two-electron states are (s = 1/2, $+d$) and (s = −1/2, $-d$). Time reversal gives transformations (s = 1/2, $+d$) → (s = −1/2, $+d$) and (s = −1/2, $-d$) → (s = 1/2, $-d$), which are not possible; therefore, T-symmetry is violated. The P-symmetry is also violated as the mirror image conversion gives the transformation (s = 1/2, $+d$) → (s = 1/2, $-d$). The PT-symmetry, however, is conserved.

The search for the eEDM should be done by measuring the Stark energy shift induced by the electric field. This is not possible however with free electrons, which are accelerated by the electric field. Therefore, atoms or molecules having non-zero electron spin are used for the measurement (figure 4.15). When the magnetic and electric fields are zero, there is no energy difference between (s = 1/2, $+d$) and (s = −1/2, $-d$) states. Given a magnetic field B, there are two Zeeman energy shifts with coefficients $+a$ and $-a$; the frequency associated with this transition between the states is

$$f_B = 2aB/h. \tag{4.3}$$

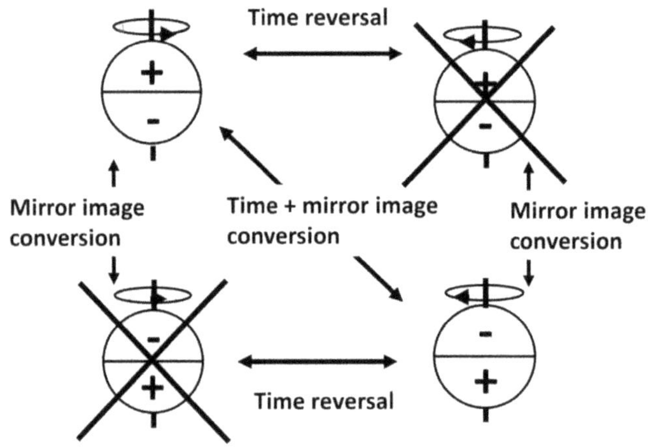

Figure 4.14. Existence of the eEDM as evidence for the violation of T- and P-symmetries.

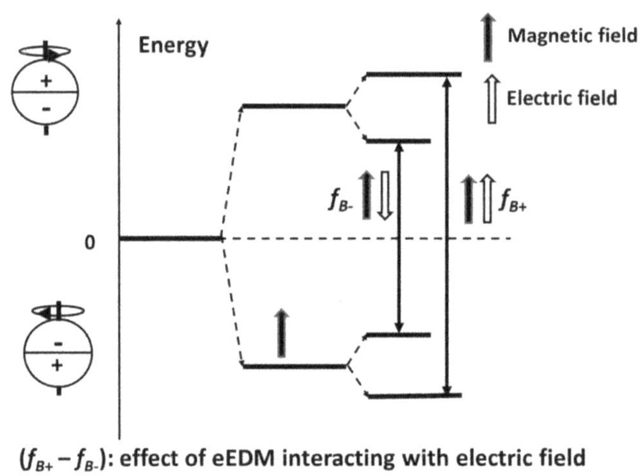

Figure 4.15. Measurement of eEDM in atoms or molecules having non-zero electron spin which provide magnetic and electric fields, simultaneously (parallel or anti-parallel).

When an electric field E is also applied parallel to the magnetic field, a Stark energy shift appears and the frequency corresponding to this transition becomes

$$f_{B+} = 2(aB + dE)/h. \tag{4.4}$$

When the electric field is applied anti-parallel to the magnetic field, the transition frequency is

$$f_{B-} = 2(aB - dE)/h. \tag{4.5}$$

Measuring the difference between these two transition frequencies,

$$f_{B+} - f_{B-} = 4dE/h, \tag{4.6}$$

yields a value for the eEDM. Here, a magnetic field is required to be applied to fix the direction of the electron spin and eEDM.

Is the eEDM detectable? The estimation from the Kobayashi–Maskawa theory is 10^{-38} e cm [16], which is not detectable. However, the CP-symmetry violation determined by this theory is not enough to account for a particle-dominant Universe and a range of values of 10^{-26}–10^{-30} e cm is estimated. The detection with electric field intensities of 10^{5} V cm^{-1} (highest electric field artificially generated) is not realistic. However, there is a possible means using electrons in atoms or molecules that experience an internal electric field strength of order 10^{10}–10^{11} V cm^{-1} as shown below. The average electric field applied to electrons in atoms or molecules is zero in the laboratory frame, otherwise they would be accelerated. However, electrons moving with high velocity also experience a magnetic field from relativistic effects and the forces from the internal electric field and magnetic field may be balanced. Therefore, electrons in atoms or molecules having heavy nuclei can experience internal electric fields higher than 10^{10} V cm^{-1}. The direction of the internal electric field is random, but it can be aligned using an external electric field (figure 4.16). We believe that the applied electric field is amplified by several orders from the internal electric field.

The requirements in the search for eEDM are:

1. High internal electric field, therefore, atoms or molecules must have heavy nuclei;
2. Advantages for precision measurements, therefore, atoms or molecules with low kinetic energy are advantageous;
3. Strong polarizability using a small external electric field to align the direction of the internal electric field.

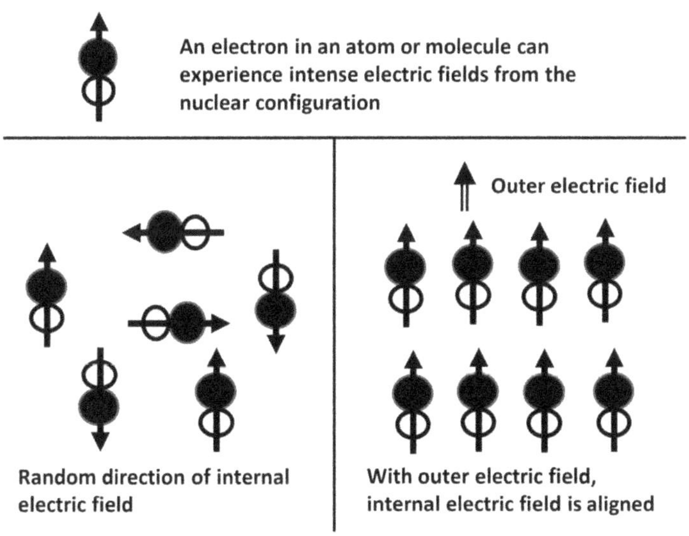

Figure 4.16. Alignment of the internal electric field.

Laser-cooled atoms hold advantages for point 2, whereas molecules with large dipole moments hold advantages for point 3. Cold polar molecules having heavy nuclei are most advantageous but have been difficult to obtain.

In 2011, a group at Imperial College London obtained an upper limit for the eEDM of 1.05×10^{-27} e cm using YbF molecules in a thermal beam [21]. In 2014, a joint group from Harvard and Yale obtained an upper limit for the eEDM of 0.87×10^{-28} e cm from a measurement using ThO molecules cooled by the vapor of liquid He [22]. To search for a non-zero value of the eEDM, the measurement accuracy of the transition frequency between two states should be further improved so that the term in equation (4.6) can be determined with an accuracy lower than 10^{-5} Hz. Recently, a group at Imperial College London succeeded in decelerating YbF molecules by laser cooling (see appendix A) [23], and a measurement uncertainty much lower than that reported in [24, 25] is expected to be obtained in the near future.

T-symmetry violation would also be demonstrated by the presence of EDMs in other particles. The fundamentals of the measurement (figure 4.15, equations (4.3)–(4.6)) are also common to other particles, such as protons or neutrons. Their estimated EDMs are several orders larger than that for an electron. However, the relativistic effect is negligibly small for heavy particles and it is difficult to have a sufficiently high internal electric field. The measurement for protons is not realistic because they are accelerated by the electric fields. Measurements of the neutron EDM with the Larmor procession giving an electric field parallel and anti-parallel direction to the magnetic field has been performed. The upper limit of the neutron EDM is estimated to be 3×10^{-26} e cm [24]. A nuclear EDM has been measured using atoms without electron spin. There is a possibility to search for an EDM in a nucleus for which the internal electric charge distribution is localized (Schiff moment). The upper limits for the ^{199}Hg and ^{129}Xe nuclei are estimated to be 3.1×10^{-29} e cm [25] and 7×10^{-27} e cm [26], respectively.

4.4 Search for the variation in the fundamental constants

Physical laws have been established with many fundamental constants, such as the speed of light c, the unit of electric charge e, and the constant of universal gravitation G. With different values of the fundamental constants, the appearance of the Universe would be quite different. For example, with a larger value of the fine structure constant $\alpha = e^2/2\varepsilon_0 hc = 0.007297$ (a parameter characterizing the strength of the electromagnetic force to the strong nuclear force) atoms with heavy nuclear could not exist because of the repulsive force between protons. With smaller values of α, molecular bonding is not possible. We would not exist with any other value of α. The combination of suitable fundamental constants is another of the mysteries of physics. If fundamental constants have a dependence on time and position, we may understand that we are living in an epoch with suitable combinations of fundamental constants; that they arose by chance is not required. In 1937, Dirac mentioned for the first time the possibility of time-varying fundamental constants [27]. However, the change in variation at that time was expected to be too small to detect with the measurement accuracies at that time. Currently, some transition frequencies can be measured with uncertainties lower than 10^{-16} and the search for such variations has become an active topic of investigation.

When there is a variation in a fundamental constant X, there is a variation in the transition frequencies. The variation in X (ΔX) is determined from the variation in the ratio of two transition frequencies $f_1(\propto X^{\lambda_1})$ and $f_2(\propto X^{\lambda_2})$,

$$\frac{\Delta(f_1/f_2)}{(f_1/f_2)} = (\lambda_1 - \lambda_2)\frac{\Delta X}{X}. \tag{4.7}$$

A non-zero $(\Delta X/X)$ is detected when $\Delta(f_1/f_2) > \delta(f_1/f_2)$, where $\delta(f_1/f_2)$ is the measurement uncertainty in (f_1/f_2). The minimum detectable value of $(\Delta X/X)$ is

$$\left(\frac{\Delta X}{X}\right)_{\min} = \frac{1}{\lambda_1 - \lambda_2}\frac{\delta(f_1/f_2)}{(f_1/f_2)}. \tag{4.8}$$

In the selection of the transition frequencies to monitor the variation in X, we have a choice between a high value in the sensitivity parameter λ or an attainable measurement accuracy. Two methods are possible in the search for the variation in the fundamental constants. One is a comparison of (f_1/f_2) values for a quasar and for Earth to see the variations over a long period. The appearance of a quasar at a distance of 10 billion light years signifies a time 10 billion years ago. With observations from quasars in different directions, we can distinguish between spatial and temporal variations. The problem with this method is the limited measurement accuracies of (f_1/f_2) for quasars. The other method involves finding a variation over several years in precision measurements of (f_1/f_2) in a laboratory. As shown below, a measurement with an uncertainty less than 10^{-16} is required for this method. We however cannot distinguish between a spatial variation and a temporal one, because the solar system moves in the Universe with a velocity of 240 km s^{-1} (one order faster than Earth's orbital velocity).

4.4.1 Variation in the fine structure constant

Estimating the sensitivity parameter of the fine structure constant α is not easy. Here a simple example is shown for the 1S–2S transition frequency of hydrogen-like ions ($M^{(Z-1)+}$). From the Schrödinger equation, we have

$$f_{1S-2S}(M^{(Z-1)+}) = p\alpha Z^2. \tag{4.9}$$

Using the Dirac equation, relativistic effects are included and equation (4.9) is revised, under an approximation, to

$$f_{1S-2S}(M^{(Z-1)+}) = p\alpha Z^2 + q\alpha^2 Z^4. \tag{4.10}$$

The second term is negligibly small for $f_{1S-2S}(H)$. The ratio between the transition frequencies of $M^{(Z-1)+}$ ion and H atom $R(M^{(Z-1)+}/H)$ is given by

$$R(M^{(Z-1)+}/H) = \frac{f_{1S-2S}(M^{(Z-1)+})}{f_{1S-2S}(H)} = Z^2 + \frac{q\alpha^2 Z^4}{p}, \tag{4.11}$$

Table 4.1. List of the variation in the fine structure constant α observed in quasar measurements. The subscripts 'stat' and 'syst' denote the statistic and systematic uncertainties, respectively.

Quasars	$\Delta\alpha/\alpha$	
Quasars at 12 billion light years (redshift z = 2.0–3.7)	$(-5.3 \pm 7.2) \times 10^{-6}$	[28]
Twelve independent quasars (10–11 billion light years, redshift z = 1.0–2.4)	$(0.4 \pm 1.4_{\text{stat}} \pm 0.9_{\text{syst}}) \times 10^{-6}$	[29]
HE0515-4414 (10 billion light years redshift z = 1)	$(-1.42 \pm 0.55_{\text{stat}} \pm 0.65_{\text{syst}}) \times 10^{-6}$	[29]

for which the second term is non-negligible for large Z. The variation in α is sought from the variation in $R(\mathrm{M}^{(Z-1)+}/\mathrm{H})$. The transition frequencies of heavy atoms depend quite sensitively on α. Different results from astronomical results have been reported and there is no clear conclusion if there is a variation in α. Recent results are shown in table 4.1. Measurements in the laboratories were also performed by comparing different atomic transition frequencies (table 4.2). The transition frequencies of heavy nuclei reflect the repulsion force between protons; therefore, they are very sensitive to the variation in α. From the ratio of the transition frequency in the ^{299}Th nucleus (section 3.7) to the $^1S_0$–$^3P_0$ transition frequency of $^{27}Al^+$ ion or ^{87}Sr atom, a variation in α of 10^{-19} seems to be detectable.

4.4.2 Variation in the proton-to-electron mass ratio

If there is variation in the fine structure constant α, there should also be variation in the proton-to-electron mass ratio μ $(= m_p/m_e)$ with a ratio larger than that for an α factor of 28–40 [35]. The proton mass m_p is 45 times larger than the total mass of the constituent particles (two up-quarks and one down-quark). Therefore, the proton mass is dominated mainly by the binding energy, which changes with the variation in the electromagnetic force between quarks ($\propto\alpha$). By comparing the variation in α and μ, we can obtain useful information for Grand Unification Theory.

A problem arises in that the atomic transition frequencies in the optical region have very low sensitivity on μ ($\lambda < 10^{-4}$). From a comparison between the Cs hyperfine transition frequency ($\propto\alpha^2/\mu$) and the optical transition frequency, $(d\mu/dt)/\mu$ was found to be $(-0.5 \pm 1.6) \times 10^{-16}$ yr^{-1} [32] and $(0.2 \pm 1.1) \times 10^{-16}$ yr^{-1} [33], respectively. It is not realistic to measure the Cs hyperfine transition frequency to an uncertainty lower than 10^{-16} (section 3.4) and detecting the variation in μ with this method appears difficult.

Molecular rotational and vibrational transition frequencies, given by the nuclear motion, have sensitivities to variations in μ; see below. The bonding length between atoms is proportional to the Bohr radius a_B, which is inversely proportional to the electron mass m_e. The frequency standard is given by an atomic transition frequency standard,

$$f_e \propto e^2/a_B \propto m_e. \tag{4.12}$$

Table 4.2. Measurement of the variation in fine structure constant α in laboratories.

Transitions	$(d\alpha/dt)/\alpha$ (yr^{-1})	
$^{199}Hg^+$ $^2S_{1/2}$–$^2D_{5/2}$ and $^{27}Al^+$ $^1S_0$–$^3P_0$ ($\lambda = -3.2$) ($\lambda = 0.008$)	$(-1.6 \pm 2.3) \times 10^{-17}$	[31]
$^{171}Yb^+$ $^2S_{1/2}$–$^2F_{7/2}$ and $^{171}Yb^+$ $^2S_{1/2}$–$^2D_{3/2}$ ($\lambda = -6.0$) ($\lambda = 1.0$)	$(-2.0 \pm 2.0) \times 10^{-17}$	[32]
	$(-0.5 \pm 1.6) \times 10^{-17}$	[33]
$^{171}Yb^+$ $^2S_{1/2}$–$^2F_{7/2}$ and ^{87}Sr $^1S_0$–$^3P_0$ ($\lambda = -6.0$) ($\lambda = 0.06$)	$(-4.3 \pm 2.5) \times 10^{-18}$	[34]

The $J \rightarrow J+1$ rotational transition frequency f_R (J: rotational quantum number) is given by

$$f_R = 2(h/4\pi^2 I)(J+1), \tag{4.13}$$

where I is the moment of inertia and $(h/4\pi^2 I)$ the rotational constant. Considering

$$I \propto m_p a_B{}^2 \propto m_p/m_e{}^2, \tag{4.14}$$

the dependence of f_R on μ is given by

$$\frac{f_R}{f_e} \propto \mu^{-1}, \tag{4.15}$$

although it is not strictly valid because of the centrifugal force. The $v \rightarrow v+1$ vibrational transition frequency f_v (v: vibrational quantum number) is given by the harmonic vibrational frequency. The vibrational transition gives the change in the electronic energy with the change in interatomic distance. Therefore,

$$\Delta f_e \propto \frac{1}{a_B} \propto m_p f_v^2 a_B{}^2,$$

$$f_v^2 \propto \frac{1}{m_p a_B{}^3} \propto \frac{m_e{}^3}{m_p},$$

$$\frac{f_v}{f_e} \propto \mu^{-1/2}. \tag{4.16}$$

Equation (4.16) is also not strictly valid because the vibrational potential is not strictly harmonic (see exercise 4.2 in chapter 4). Molecular transition frequencies between splitting states given by the tunneling effect of atoms have higher sensitivities to the variation in μ For example, $\lambda = -4.3$ with ammonia inversion transition [36]. Considering also the attainable measurement accuracy, a pure vibrational transition frequency (a vibrational transition keeping all angular momentum quantum states constant) offers more advantages in being measurable with the lowest uncertainty and in seeking a variation in μ (section 3.6). Measurements of the vibrational transition frequency of SF_6 molecules yielding an uncertainty of order 10^{-14} were performed for this purpose [37]. This measurement was performed with a thermal molecular beam. Using cold molecular ions, the vibrational transition frequencies can be measured with much lower uncertainties [38–40]. The $^1S_0$–$^3P_0$ transition frequencies of $^{27}Al^+$ or ^{87}Sr atom are good candidates as references because of the low frequency uncertainties (of order 10^{-18}) and low sensitivities in the variations in μ and α.

From the H_2 molecular transition frequencies in the quasar source J2123-005 (redshift $z = 2.059$, 12 billion light years away) observed by two different telescopes, $\Delta\mu/\mu=(8.3\pm3.6_{stat} \pm 2.2_{syst}) \times 10^{-6}$ and $(5.6\pm5.5_{stat} \pm 2.9_{syst}) \times 10^{-6}$ were obtained [41]. A value of $\Delta\mu/\mu=(0 \pm 1) \times 10^{-7}$ was obtained with measurements of the CH_3OH transition frequency in a quasar PKS1830-211 (redshift $z = 0.89$, 80 light years away) [42].

4.5 Search for symmetry violation in chiral molecules

Chiral molecules are molecules that are non-superimposable on their mirror images. The mirror images of a chiral molecules are called optical isomers (figure 4.17). Individual optical isomers are designated as being either right- or left-handed.

The mystery is that the abundances of right- and left-handed optical isomers are not always equal, indicating some violation of chiral symmetry. The reason is believed to stem from the P-symmetry violation induced by the weak nuclear force [43].

The atomic bonding of right- and left-handed optical isomers are separated by a potential barrier (figure 4.18). The transformation between the two optical isomers is possible by the tunneling effect (a quantum phenomenon which makes it possible for an object of low energy to pass through a high potential barrier). If the structures of the atomic bonding are symmetric for both optical isomers, the transformation rate is equal for both directions. However, if there is a slight asymmetry, the transformation rate from the higher to the lower energy state is much higher than that for

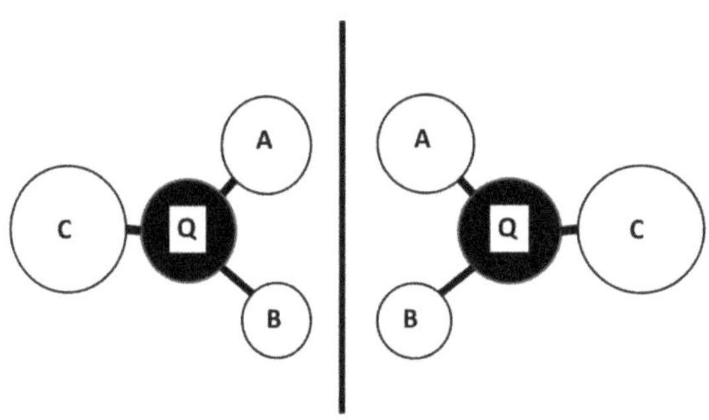

Figure 4.17. Optical isomers of chiral molecules.

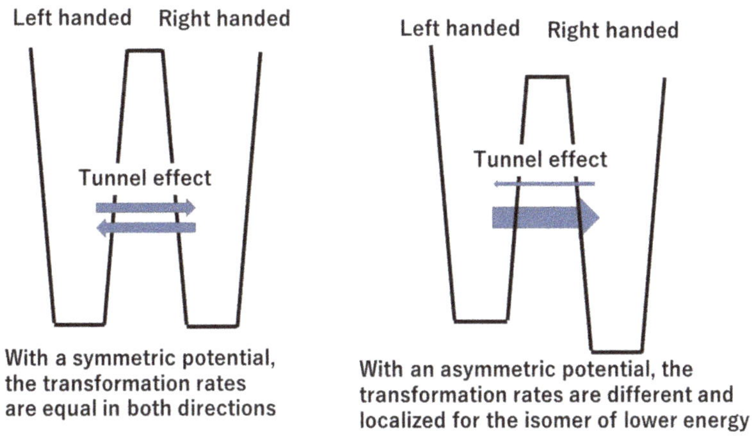

Figure 4.18. Transformation between an optical isomer pair passing through a potential barrier.

the reverse direction. Therefore, the abundance of the optical isomer with the lower energy state is much larger than that for the other.

To confirm the explanation presented above, the slight difference in the vibrational transition frequencies must be detectable. The localized abundance of the optical isomers is consistent with the theoretical estimate if there is a difference in the vibrational transition frequencies of about 0.1 Hz (10^{-14}). This accuracy is not easily attained with the transition frequencies of polyatomic molecules at room temperature. Several groups are in the process of developing experimental devices for this purpose [44]. Laser cooling of polyatomic molecules is a first step to detect the asymmetry in the different optical isomers. Recently, a Harvard group succeeded in decelerating SrOH molecules by laser cooling [45]. The precision measurement of the molecular transition frequency advances not only physics but also molecular biology because many biologically active molecules are chiral, including amino acids and sugars. Amino acids are mostly left-handed and sugars are mostly right-handed. With the equal abundance of optical isomers, the appearance of organisms might have been something different.

4.6 Role of atomic clocks for astronomy

Fundamental to astronomy is the observations of the position and motion of stars. As depicted in figure 4.19, the positions of stars are determined by the displacement seen when observed from two or more telescopes at distant places, much like the binocular vision of humans and animals. The direction of observation changes as time elapses, and therefore, simultaneity of observation is a very important factor in providing measurement uncertainty. Atomic clocks play a very important role in guaranteeing simultaneity with high accuracy.

Analyzing the radiation emanating from stars, we see there are particularly strong or weak intensities at specific frequencies (figure 4.20). From this result, we can discover the material composition of celestial objects. The transition frequencies observed from stars are shifted because of the Doppler effect. From the shifts in the

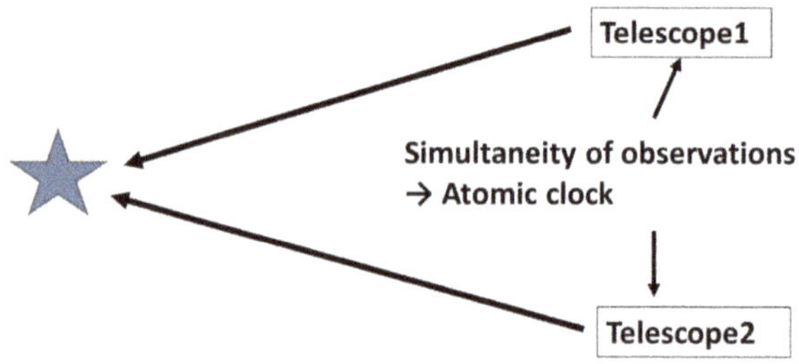

Figure 4.19. Measurement of the position of a star using two telescopes.

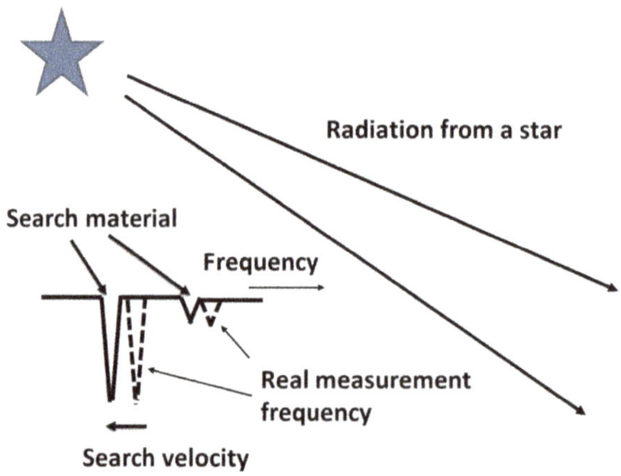

Figure 4.20. Measurement of the material and velocity of a star from the frequency distribution of radiation.

transition frequencies, we can determine the velocity of motion of the stars. The accuracy obtained for the shifts in the observed transition frequencies determines the accuracy of the information regarding the motion of stars.

From the precision measurement of positions and velocities of stars, we can develop evolutionary details regarding the Universe.

4.6.1 History of the Universe

The theory of general relativity indicated that the Universe is not stable but expanding. This conclusion was surprising at that time (Einstein himself could not accept it at first), but the fact was confirmed when Hubble's law was discovered in 1929 [46]. This law stated that the objects observed in deep space are moving away from Earth with a velocity approximately proportional to their distance from Earth. This law implies that the Universe was born 13.8 billion years ago from an explosion called the 'Big Bang' (figure 4.21). This idea was confirmed with the discovery of the cosmic microwave background (CMB) in 1965 [47]. The CMB was observed to be uniform to an accuracy of 10^{-5} in all directions with a frequency distribution corresponding to blackbody radiation with an absolute temperature of 2.725 K. The uniform temperature shows that the Universe was a single object in the past.

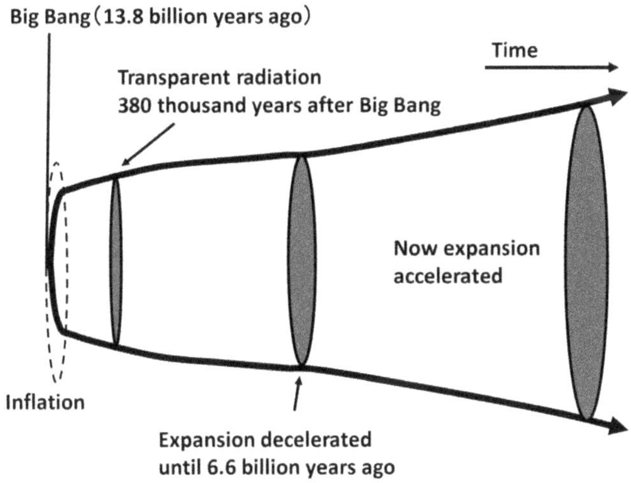

Figure 4.21. History of the Universe after the Big Bang.

The temperature of the Universe during the Big Bang was unimaginably high and the subsequent speed of expansion (called inflation) was much faster than the speed of light. As the Universe expanded, the temperature decreased. After 380 000 years, the temperature was around 3000 K, commencing a period when atoms formed from nuclei and electrons. After this period, the Universe became transparent to electromagnetic waves (the transparent radiation of the Universe) and the CMB is understood to be the radiation from this period. The CMB observed on Earth is redshifted from the continued expansion of the Universe.

After establishing details of the Big Bang, astronomers have been eager to observe galaxies in distant places to understand the appearance of the Universe soon after its dawning.

4.6.2 Dark energy

Under gravitational interactions, one can conceivably accept that the velocity of expansion decreases. However, precision measurements of the positions and velocities of galaxies show that the expansion of Universe has been accelerating for the past 6.62 billion years, having decelerated until then. The force causing this acceleration is called 'dark energy' [48]. It is not known if dark energy is the same force that produced inflation soon after the Big Bang.

The true nature of dark energy is an added mystery of modern physics. To seek further information about dark energy, detailed observations of new stars at great distances is required.

The precise measurement of the temperature distribution of the CMB also gives useful information concerning dark energy. The temperature distribution of the CMB was believed to be homogeneous until the 1980s, but with improved accuracies the fluctuation at levels of 10^{-5} persisted [49]. This fluctuation derives from the nonhomogeneity during the dawn of the Universe. The general understanding is that no galaxies or stars could appear if the CMB were perfectly homogeneous.

4.6.3 Dark matter

In astronomy, there is a further mystery in the form of dark matter. Dark matter is a substance that interacts gravitationally but not electromagnetically. For this reason, we cannot observe it directly.

The existence of dark matter is determined from the velocity of H atoms at the edges of galaxies [50]. The measured velocity is much higher than expected and the gravitational force required to balance the centrifugal force is ten times larger than that given by the matter within each galaxy (figure 4.22). With no observed electromagnetic spectra, the nature of this unknown substance remains a mystery.

In the 1980s, the distribution of groups of galaxies was mapped and a bubble-like structure of the Universe appeared (figure 4.23) [51], along with areas called voids, where there are no galaxies. The sizes of these voids are larger than 100 million light years. This structure shows that the mass of the Universe is dominated more by dark matter than matter. Recently, a galaxy without dark matter NGC1052-DF_2 was observed [52] and a new model may be needed in the near future to show that galaxies can form without dark matter.

While we cannot observe dark matter directly, its distribution can be measured using gravitational lensing (the bending of the optical path by gravity) which was shown by the theory of general relativity (figure 4.24) [53]. Moreover, gravitational lensing by dark matter makes it possible to observe a single star that is nine billion light years away [54].

That the true nature of dark matter is not known is in itself not strange as there are many elementary particles (specifically, quarks) whose existence is theoretically predicted but not isolated. Recently, the variation in fundamental constants

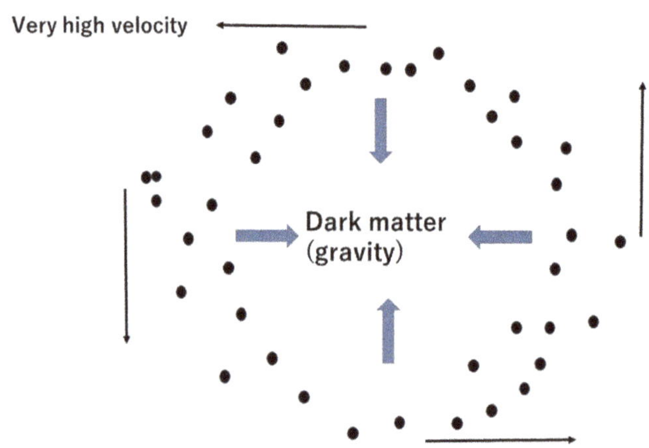

Figure 4.22. H atoms at the end of galaxies are moving very fast and the gravitational force from dark matter is required to balance the centrifugal force.

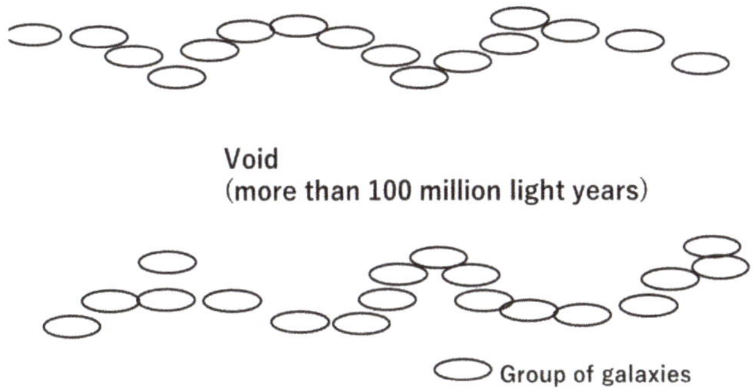

Figure 4.23. The distribution of groups of galaxies appear to have a bubble-like structure.

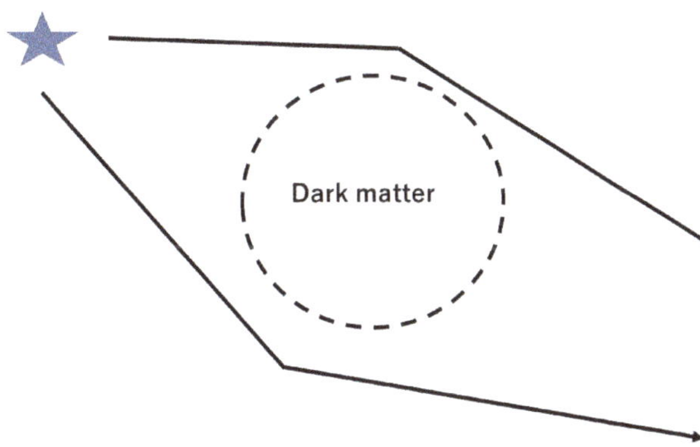

Figure 4.24. Measurement of the distribution of dark matter using a gravitational lens.

(section 4.4) as determined by the density of dark matter has become doctrine among many [55].

4.7 Quantum electrodynamics and proton size

The energy structure of the hydrogen atom has been solved using the Schrödinger and Dirac equations. From the Schrödinger equation, the $2S_{1/2}$, $2P_{1/2}$, and $2P_{3/2}$ states should have the same energy. From the Dirac equation, the $2P_{3/2}$ states have higher energy than the other two states because of relativistic effects (the transition frequency being 10.9 MHz), with the $2S_{1/2}$ and $2P_{1/2}$ states having the same energy. However, an energy splitting between the $2S_{1/2}$ and $2P_{1/2}$ states with a transition frequency of 1.04 MHz was discovered (figure 4.25) [56]. This frequency shift, called the 'Lamb shift', is explained from the interaction between the vacuum energy fluctuations and the hydrogen electron in different orbits. The Lamb shift played a significant role in developing the then-emerging field of 'quantum electrodynamics' in light of the vacuum energy fluctuation. The ratio of the Lamb shift to the 1S—2S transition frequency is 4.4×10^{-10}. Precision measurements of transition frequencies remain very much to the fore in research on quantum electrodynamics.

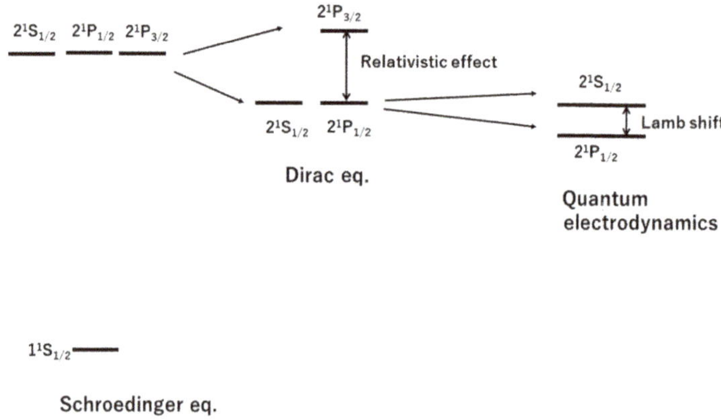

Figure 4.25. Schematic of the energy structure of the H atom obtained using the Schrödinger and Dirac equations and quantum electrodynamics.

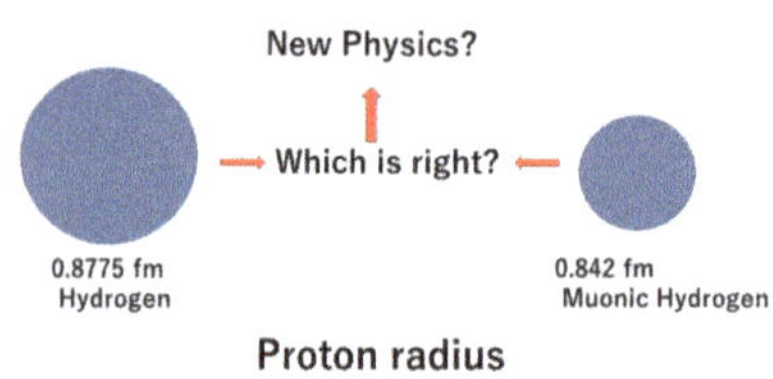

Figure 4.26. Proton radius puzzle.

To obtain more details of the Lamb shift, measurements of hydrogenic ions (nucleus with Z^+ + one electron) have been performed [57] because the Lamb shift is proportional to Z^4.

With quantum electrodynamics, the energy levels of the hydrogen atom can be calculated with high accuracy. The remaining source of discrepancies between experiment and calculation arises from the finite size of the proton. Estimated to be 0.8775 ± 0.005 fm [58], its accuracy was expected to improve with measurements of muonic hydrogen atom (proton + muon) because the muon's orbital radius is much smaller than that of the electron and therefore the influence of proton size is more significant. However, the result obtained was 0.842 ± 0.001 fm [59] (figure 4.26). The discrepancy in the results is larger than five times the measurement uncertainty and thus has become a mystery of modern physics called the 'proton radius puzzle'. Resolving the puzzle may provide a chance to develop new physics, including extra dimensions. Again, the precision measurements of the transition frequencies of the hydrogen atom have initiated not only the development of quantum electrodynamics, but also a window into quantum chromodynamics.

Quantum electrodynamics is confirmed not only from the Lamb shift, but also from another phenomenon. The g-factor of the electron spin g_S (section 2.2) is from the Dirac equation exactly 2. However, precision measurements give a value of 2.002 319 with an uncertainty of 10^{-13} [60]. A non-zero value of (g_S–2) has been explained within quantum electrodynamics [61]. Precise measurements of (g_S–2) provide details of the internal structure of the electron. Currently, the electron radius is estimated to be smaller than 1×10^{-18} m. Uncertainties in both measurement and calculation have been improved and the issue now remains in the balance; either quantum electrodynamics explains everything about the electron or new physics is required. For the *g*-factor of the muon spin, there is a discrepancy between measurement and calculation of about 2×10^{-10}, whereas the measurement and calculation uncertainties are 4×10^{-12}.

Exercises

4.1 We assume that the electron electric dipole moment d is 1×10^{-30} e cm and the internal electric field E is 10^{11} V cm^{-1}. With the method of section 3.3.3, the detection of the electric dipole moment is given by the frequency $4dE/h$ (equation (4.6)). Estimate this value. The unit charge is $e = 1.6 \times 10^{-19}$ C and Planck's constant is $h = 6.6 \times 10^{-34}$ J Hz^{-1}. We shall see later how difficult it is to detect.

4.2 The $v = 0 \rightarrow v'$ vibrational transition frequency is given by $f_c = f_v v' - xfv'(v' + 1)$, taking also anharmonic terms into account. f_v and xf are proportional to $\mu^{-1/2}$ and μ^{-1}, respectively, (μ: proton-to-electron mass ratio). Obtain $\lambda = (\Delta f_c/f_c)/(\Delta\mu/\mu)$.

Solutions

4.1 9.7×10^{-5} Hz.

4.2 From $\frac{df_c}{d\mu} = -\frac{1}{2}\frac{f_v}{\mu}v' + \frac{xf}{\mu}v'(v'+1)$, $\lambda = -\frac{1}{2} + \frac{xf}{f_v}(v'+1)$ is obtained assuming $f_v \gg xf$.

References

[1] Mungall A G 1971 *Metrologia* **7** 49

[2] Chou C W 2010 *Science* **329** 1630

[3] Briatore L and Leschiutta S 1977 *Nuovo Cim.* **37B** 219

[4] Taylor J H and Weisberg J M 1982 *Astrophys. J.* **253** 908

[5] Abbott B P *et al* 2016 *Phys. Rev. Lett.* **116** 061102

[6] Abbott B P *et al* 2016 *Phys. Rev. Lett.* **119** 141101

[7] Shappee B J *et al* 2017 *Science* **358** 1574

[8] Yiu Y 2017 The past, present and future of gravitational waves *Inside Science*

[9] Salumbides *et al* 2015 ArXiv:1502.02838

[10] Ubachs W 2015 arXiv: 1511:00985v1

[11] Atkins P W 1974 *Quanta: A Handbook of Concepts* (Oxford: Clarendon), p 52

[12] Schwinger J 1951 *Phys. Rev.* **82** 914

[13] Wu C S *et al* 1957 *Phys. Rev.* **105** 1413

[14] Christenson J H *et al* 1964 *Phys. Rev. Lett.* **13** 138

[15] Alavi-Harati A *et al* 1999 *Phys. Rev. Lett.* **83** 22

[16] Lees J P *et al* 2012 *Phys. Rev. Lett.* **109** 211801

[17] Kobayashsi M and Maskawa T 1973 *Prog. Theor. Phys.* **49** 652

[18] Hori M *et al* 2011 *Nature* **475** 484

[19] Ulmer S *et al Nature* **524** 196
[20] Ahmadi M 2017 *Nature* **541** 506
[21] Hudson J J *et al* 2011 *Nature* **473** 493
[22] Baron J *et al* 2014 *Science* **343** 269
[23] Lim J *et al* 2018 *Phys. Rev. Lett.* **120** 123201
[24] Afach S *et al* 2015 *Phys. Rev.* D **92** 092003
[25] Griffith W C *et al* 2009 *Phys. Rev. Lett.* **102** 101601
[26] Rosenberry M A *et al* 2001 *Phys. Rev. Lett.* **86** 22
[27] Dirac P A 1937 *Nature* **139** 323
[28] Le T D 2015 *Astrophysics* **58** 14
[29] Murphy M T *et al* 2016 *Mon. Not. R. Astron. Soc.* **461** 2461
[30] Kotus S M *et al* 2017 *Mon. Not. R. Astron. Soc.* **464** 3679
[31] Rosenband T *et al* 2008 *Science* **319** 1808
[32] Huntemann N *et al* 2014 *Phys. Rev. Lett.* **113** 210802
[33] Godun R M *et al* 2014 *Phys. Rev. Lett.* **113** 210801
[34] Peik E 2017 *1st North American Conf. on Trapped Ions* Boulder, CO, 14–18 August
[35] Calmet X and Fritzsch H 2002 *Eur. Phys. J.* D **24** 639
[36] Bethlem H L *et al* 2008 *Eur. Phys. J. Spec. Top.* **163** 55
[37] Shelkovnikov A *et al* 2008 *Phys. Rev. Lett.* **100** 150801
[38] Kajita M and Moriwaki Y 2009 *J. Phys. B: At. Mol. Opt.* **42** 154022
[39] Kajita M *et al* 2014 *Phys. Rev.* A **89** 032509
[40] Kajita M 2017 *Phys. Rev.* A **95** 023418
[41] van Weerdenburg F *et al* 2011 *Phys. Rev. Lett.* **106** 180802
[42] Bagdonaite J *et al* 2013 *Science* **339** 46
[43] Quack M 2002 *Angew. Chem. Int. Ed.* **41** 4618
[44] Quack M 2008 *Ann. Rev. Phys. Chem.* **59** 741
[45] Kozyryef I *et al* 2017 *Phys. Rev. Lett.* **119** 133002
[46] Riess A *et al* 1998 *Astrophys. J.* **116** 1009
[47] Penzias A A and Wilson R W 1965 *Astrophys. J.* **142** 419
[48] Perlmutter S, Turner M S and While M 1999 *Phys. Rev. Lett.* **83** 670
[49] Smooth G F *et al* 1992 *Astrophys. J. Lett.* **396** L1
[50] Rubin V, Thonnard N and Ford W K Jr 1980 *Astrophys. J.* **238** 471
[51] Gregory S A and Thomson L A 1978 *Astrophys. J.* **222** 784
[52] van Dokkum P *et al* 2018 *Nature* **555** 629
[53] Aubourg E *et al* 1993 *Nature* **365** 623
[54] Rodney S A *et al* 2018 *Nat. Astron.* **2** 324
[55] Stadnik Y V and Flambaum V V 2015 *Phys. Rev. Lett.* **114** 161301
[56] Lamb W E and Retherford R C 1947 *Phys. Rev.* **72** 241
[57] Kugel H W and Murnick D E 1977 *Rep. Prog. Phys.* **40** 297
[58] Sick I and Trautmann D 2014 *Phys. Rev.* C **89** 0122010
[59] Pohl R *et al* 2010 *Nature* **466** 213
[60] Odom B *et al* 2006 *Phys. Rev. Lett.* **97** 030801
[61] Brodsky S J *et al* 2004 *Nucl. Phys.* B **703** 3

Chapter 5

Role of atomic clocks for human life

5.1 Position measurement by GPS [1]

In the 'Age of Discovery' (the fifteenth to eighteenth centuries), navigators found it necessary to measure their own positions. Latitude was measured from the height of Polaris above the horizon. However, to measure longitude, the role of the clock was very important because longitude is given by the position of the constellations at a given time of day. Improved methods of time-keeping provided more accurate positioning.

Currently, latitude, longitude, and altitude are measured by receiving radio signals from satellites, called the global positioning system (GPS). This system monitors the position of ships, trains, and airplanes. Recently, the use has expanded to car drivers and pedestrians in cities. The measurement is based on the distance between a satellite and a receiver as determined from the propagation time of the radio signal. From the distance from one satellite, the position of the receiver is determined to be on a sphere with a radius of measured distance. By measuring the distances to three satellites simultaneously, the position of the receiver can be determined by trilateration. The clocks in the satellites are atomic clocks, but a quartz clock is used for the receiver. Therefore, the accuracy of the measured distances to the satellites is not high enough and an error of 10^{-6} s in time corresponds to a positional error of 300 m. If we are able to measure distances to four satellites, the error of the receiver's clock is corrected (figure 5.1). Therefore, it is preferable to receive radio signals from at least four satellites and invoke multilateration.

Currently, there are 32 satellites (height 20 000 km) being used for GPS geo-location, and a GPS user can receive radio signals from six to ten satellites anywhere on Earth. To obtain the distance between the satellite and receiver to within an accuracy of 1 m, an accuracy of 3×10^{-9} s is required from satellite clocks. Therefore, the accuracy in determining the position depends on the precision of these clocks (figure 5.1).

doi:10.1088/978-0-7503-2124-2ch5

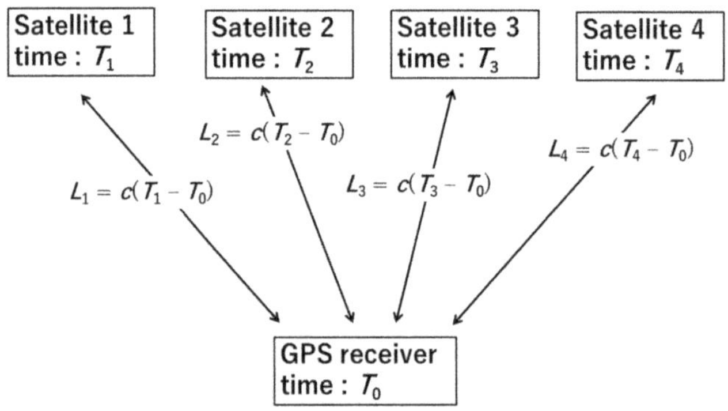

Figure 5.1. Position measurement by GPS. Distances from four satellites are calculated from the propagation time of the electromagnetic wave.

The GPS system was developed by the US for military use in 1973 and was opened for civil use after the incident involving Korean Air Flight 007 (1983). However, intentional degradation of the accuracy, called selective availability (SA), was added to civil users' times for the US military to maintain their advantage over foreign military forces. While SA was operating, the position accuracy was of order of 100 m. In 2000, the US government discontinued the use of SA. The main reason was that other countries were also launching their own satellites, and the US no longer remained the sole provider of space-based position measurements. At the moment, an accuracy of a few meters can be given with civilian receivers.

As mentioned above, radio receivers need to be linked to at least four satellites. Nevertheless, there are several additional methods to perform measurements utilizing radio signals from just three satellites. After the clock in the receiver is corrected once using radio signals from four satellites, the position measurement is possible while the receiver's clock continues to show the correct time. The period in which we can use this method is, however, not very long. With a chip-scale atomic clock (see section 3.4) mounted in a GPS receiver, this problem is overcome.

5.2 Very long baseline interferometry

As described in section 4.6, the role of the atomic clock is significant in measuring the position of stars using several telescopes. Moreover, the distance between telescopes can be measured to millimeter accuracy from the reception of radio pulses originating from quasars (figure 5.2). This system is called the very long baseline interferometry (VLBI). In one application, the variation in the distance between two different continents or islands can be monitored using VLBI. In one application, the distance between Kashima (Ibaraki, Japan) and Koganei (Tokyo, Japan) was found to be closing by 2.6 mm yr^{-1}, but soon after the 'Great East Japan Earthquake' (Tohoku, 2011) the distance opened by 55 cm [2]. In the near future, VLBI may be the most useful method to predict earthquakes or volcanic eruptions. For measurements of distances using VLBI, the accuracy of the time delay must be measured with high accuracy; for example, an error of 10^{-10} s produces an error of 3 cm, which may suffice in indicating an early warning.

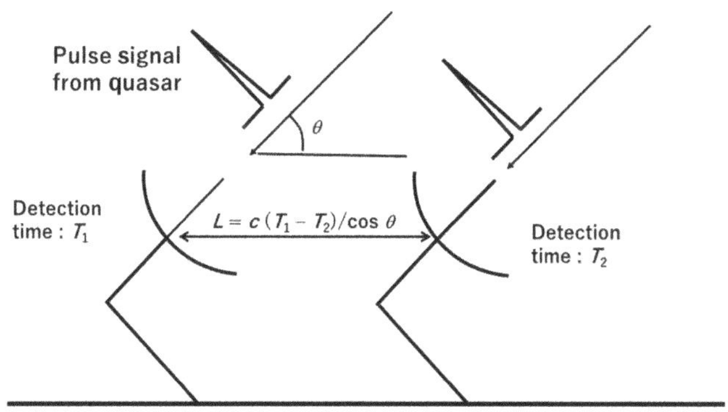

Figure 5.2. Distance measurements using VLBI.

5.3 Establishment of a national standard time and its distribution to users

The time scales created by atomic clocks are regularly compared and combined to form a stable time scale. International Atomic Time (TAI) was inaugurated at 0:00 January 1, 1958. The time was defined by Bureau International des Poids et Mesures (BIPM) using a weighted averaging of times that are broadcast by atomic clocks all over the world. However, we cannot ignore the fact that the daily lives of humans and the annual cycle of the seasons correlate with the motion of the Sun. The solar-based Universal Time (UT) must be maintained close to TAI. The Coordinate Universal Time (UTC) is the time scale obtained by inserting an additional one second (called the leap second) to the TAI, so that the difference between UTC and UT is less than 0.9 s. Currently (2018), UTC is lagging by 37 s behind TAI. Recently, there has been a debate whether to keep adjusting UTC with leap seconds to maintain synchronicity with Earth's slowing rotation—the discussion continues.

The UTC is in a sense a paper clock, which is calculated every five days and published one month later as the difference from standard times of each countries. The standard time in each country is given by atomic clocks as the real time. Physikalische Technische Bundesanstalt (Germany), National Institute of Standard Technology (NIST, US), and Systèmes de Référence Temps Espace (France) give national standard times mainly using Cs atomic clocks with an atomic fountain (section 3.4). The National Institute of Information and Communications Technology (NICT, Japan) is establishing a Japanese standard time using a maximum of 18 commercial Cs clocks and one hydrogen maser.

Recently, a new method was developed to establish the standard time using atomic clocks operating in the optical region (figure 5.3). In NICT, the real time signal of an accurate time scale was made using a hydrogen maser as a source oscillator and corrections obtained from measurements using an ^{87}Sr lattice clock (section 3.5) taken over one 3 h period per week [3]. By continuing this measurement for six months, the accuracy obtained was much higher than that obtained using the

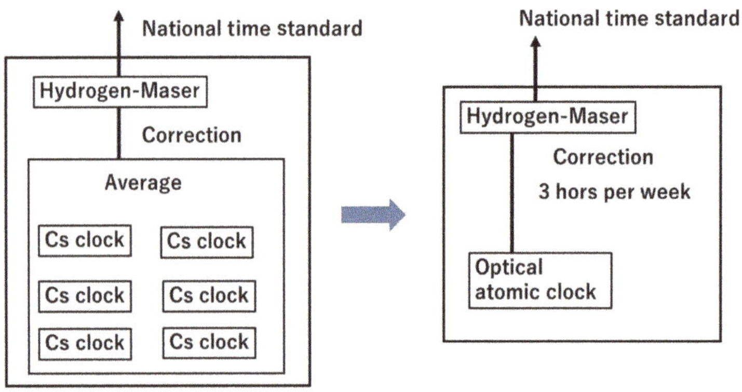

Figure 5.3. Determination of the national time standard with the present system using Cs atomic clocks and a future system using an optical atomic clock.

18 commercial Cs atomic clocks. In the near future, each national standard time is likely to be established using atomic clocks operating in the optical region.

The national standard times are distributed to users by radio signals (25 kHz–25 MHz) [4], the Internet, and telephone lines. The radio clock is a quartz clock that is corrected by a radio signal from the national standard time. The difference from the standard time is less than 0.01 s. The finite propagation time of the radio signal limits the accuracy. Some radio-clock receivers combine multiple time sources to improve their accuracy. The radio signal includes information about summer time and the leap second. The national standard time is distributed by radio signals to a wide range of other receivers such as clocks inserted in home electric applicants, cameras, and automobile clocks.

The national standard time distributed by the Internet can synchronize the time of networked computers. A high precision time signal is also supplied via telephone lines, over which time information is sent to personal computers, modems, and commercial communications software.

5.4 Contribution to communication technology

In section 3.8, the optical fiber link was described as being useful for frequency comparisons and the distribution of an accurate frequency signal. Such links are useful in communications technology. We can transfer carrier signals by laser light with wavelengths between 1530 and 1565 nm; in this C-band, power losses in the fiber are minimal. To convert different carrier signals simultaneously, each signal is transformed into laser light of a different frequencies, in an analogous way to selecting TV channels using different radio frequencies. For good signaling, there should be no frequency overlap between the different carrier signals. This is called wavelength division multiplex (WDM).

The volume of transformable information or network capacity is given by the stability in the frequency of laser light used. A device was developed that stabilized the frequency of a laser diode to the vibrational–rotational transition frequencies of acetylene (C_2H_2) molecules. The C_2H_2 transition frequencies lie in the C-band at

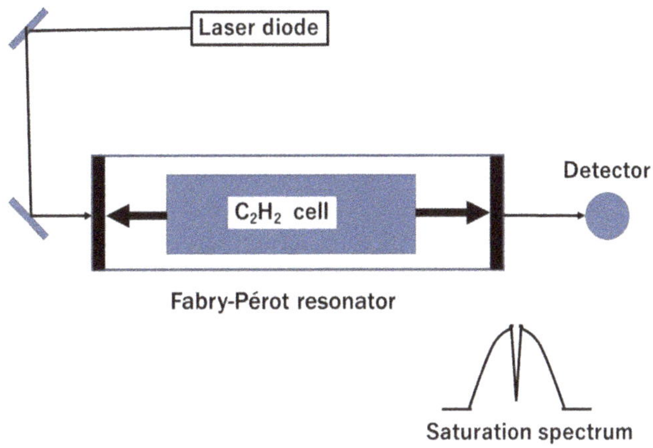

Figure 5.4. Laser frequency stabilizer using C_2H_2 absorption cell in a Fabry–Pérot resonator to observe the saturation spectrum.

193–196 THz (1530–1550 nm) with intervals of 100 GHz. To achieve high-frequency stability, using transitions with narrow linewidths of the saturated absorption lines of acetylene are preferable, but this is difficult to bring about successfully with non-polar molecular transitions. To overcome the problem, an absorption cell was developed and inserted in a Fabry–Pérot resonator, so that the power density in the cell becomes high enough to observe the saturated absorption (figure 5.4) [5]. Currently, the channel spacing for WDM is 100 GHz and the C_2H_2 transition frequency uncertainty is 10–100 kHz.

5.5 Monitoring air composition

Over recent times, the abundance of carbon dioxide (CO_2) in the atmosphere has been increasing rapidly. Within the last 200 years, it has increased from 280 ppm to 400 ppm. To monitor this abundance, eye-safe laser light (2 μm) is transmitted in the air and the absorption by CO_2 is registered. The power loss of the laser light is caused by not only CO_2 absorption but also the scattering by aerosols or clouds. For this reason, laser light with two frequency components are irradiated, one component being resonant to the CO_2 transition frequency and the other component to an off-resonant frequency. The pure absorption signal is obtained from the difference in power loss for both frequency components. With this method, the abundances of CH_4 and N_2O are also recorded (see figure 5.5).

Using the same method, the laser-light absorption by water vapor can be monitored as well. Atmospheric humidity, i.e. the density of water vapor in air, is useful to estimate rainfall percentages. From the absorption of laser light, THz waves, or microwaves, the detection of very small fractions of other constituents in the atmosphere are also possible.

The atmosphere can also be monitored for air pollution using scattered radiation of laser light. From the scattered light, we can monitor the abundance of aerosols at

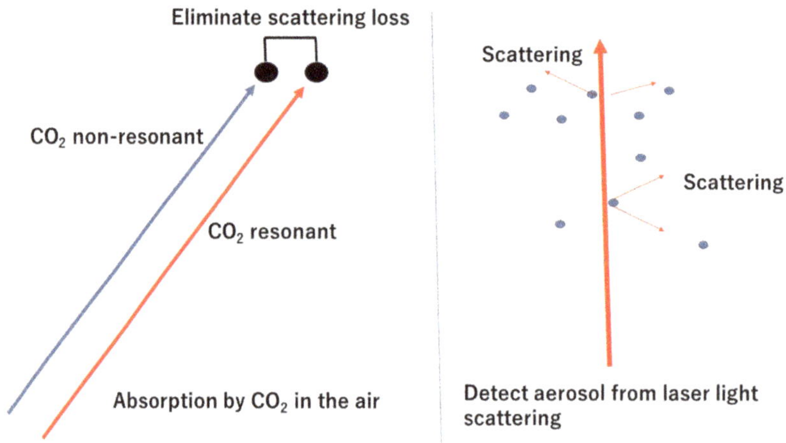

Figure 5.5. Monitoring the atmosphere by irradiating with eye-safe laser light to observe the absorption by CO_2 gas or scattering by aerosols.

different heights (see figure 5.5). Using a pulse laser, the distance to the aerosol is monitored by recording the time differences between transmission and reception.

When aerosols are moving, the frequency of scattered light is shifted by the first-order Doppler effect, and thereby wind speeds are determined. This information is very useful in estimating the trajectories of typhoons.

Ozone levels can be monitored by measuring the intensity of ultraviolet (UV) light using two frequency components: one that is absorbed by ozone and the other not. From the ratio of intensities for both frequency components, the ozone concentration is calculated. This information is important in setting UV protection policies and providing warnings regarding over-exposure.

The state of the environment can be monitored comprehensively using a range of frequencies afforded by eye-safe laser light, THz waves, microwaves, or UV light.

5.6 Temperature measurement

The definition of the absolute temperature scale is defined such that the triple point of water is 273.16 K, the uncertainty for which is 0.1 mK. The Boltzmann constant k_B has been measured either using an acoustic gas thermometer or through Doppler broadening of the transition frequency regime of molecules [7]; currently, its uncertainty is less than 10^{-6}. In November 2018, a proposal will be laid out to define Boltzmann's constant as a value without uncertainty. With this new definition, the definition of absolute temperature will no longer be valid.

With a thermal equilibrium state, the distribution at each state with energy E is proportional to $\Omega \exp(-E/k_B T)$, where Ω is the degeneracy of states. Observing the absorption of laser light at different frequencies (figure 5.6), the temperature can be determined. Here we consider the measurement of temperature using molecules in the gaseous state. From the population distributions of their states, molecules having many vibrational–rotational states offer more advantages than atoms in temperature measurements.

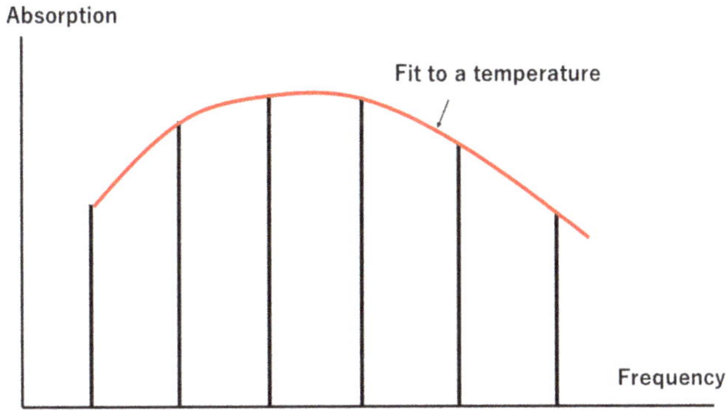

Figure 5.6. Temperature can be obtained from the distribution of each energy state, observed from the absorption of laser light at different frequencies.

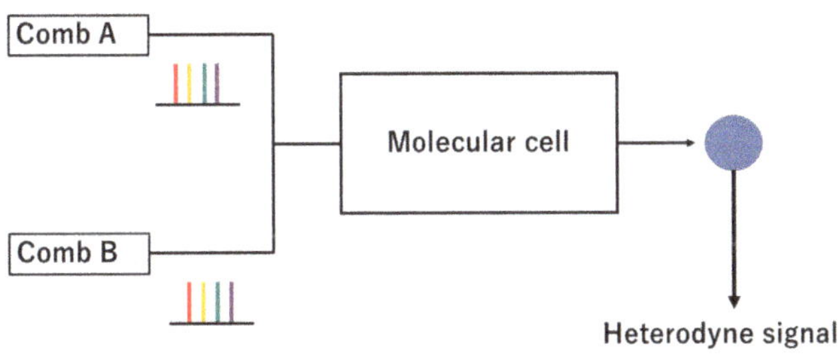

Figure 5.7. Schematic of the measurement apparatus using dual frequency combs.

This measurement is performed observing the change in the absorption signal intensities while scanning with the laser at different frequencies. It takes time to observe the many absorption lines, and the temperature can fluctuate while taking measurements.

Recently a new method using two frequency combs was developed (figure 5.7). Described in section 3.5, a frequency comb is an ultra-short pulse laser with a stable repetition frequency, and its frequency components are equally distributed at intervals of the repetition frequency. We consider frequency combs A and B with the following frequency components:

$$\text{Comb A} \quad f_A(N_c) = f_{\text{ceo}} + N_c f_{\text{rep}},$$
$$\text{Comb B} \quad f_B(N_c) = f_{\text{ceo}} + \delta f_{\text{ceo}} + N_c(f_{\text{rep}} + \delta f_{\text{rep}}).$$

When the N_cth frequency component of comb A is absorbed by the atoms or molecules, its power loss is observed as a heterodyne signal with the N_cth frequency component of comb B of frequency $\delta f_{\text{ceo}} + N_c \delta f_{\text{rep}}$. From the frequency of the heterodyne signal, we know which frequency component of the comb was absorbed. When the N_{c1}th and N_{c2}th frequency components of comb A are simultaneously absorbed, the heterodyne signal is constructed by the frequency components of $\delta f_{\text{ceo}} + N_{c1} \delta f_{\text{rep}}$ and $\delta f_{\text{ceo}} + N_{c2}\, \delta f_{\text{rep}}$, which can be resolved by mass spectrometry. With this method, many transitions can be observed with a single measurement.

The temperature calculated from the distribution of rotational states of C_2H_2 molecules was achieved within an accuracy of 1 K [8]. Non-polar molecules provide some advantages over polar molecules in temperature measurements because their absorption intensity is proportional to the number of molecules in each energy state (a small saturation effect).

The measurement of many molecular transition frequencies using dual frequency combs is also useful for the analysis of the energy structure of polyatomic molecules. Moreover, this method is useful in revealing the chemical structure of unknown materials.

Exercises

5.1 For linear molecules (such as C_2H_2), the rotational states are given by quantum number J. The degeneracy of the states is $\Omega = (2J + 1)$ and the energy of the Jth level is $E = hB_v\, J(J + 1)$ (B_v is a rotational constant). Obtain the rotational state, for which the population is maximum at absolute temperature T.

Solutions

5.1 The population is proportional to $(2J + 1)\exp\left[-\frac{hB_v J(J+1)}{k_B T}\right]$
This population is maximum with

$$J = \frac{1}{2}\left(\sqrt{\frac{2k_B T}{hB_v}} - 1\right).$$

References

[1] Federal Aviation Administration 2014 Satellite Navigation—Global Positioning System (GPS)

[2] National Institute of Information and Communications Technology www.spacegeodesy.go.jp/vlbi/image/timeseries_Ts.png (in Japanese)

[3] Hachisu H *et al* 2018 *Sci. Rep.* **8** 4243

[4] Wikipedia 2018 Radio clock

[5] Onae A *et al* 1999 *IEEE IM* **48** 563

[6] NICT Remote Sensing Laboratory

[7] Fischer J 2016 *Philos. Trans. R. Soc.* A **374** 20150038

[8] Shimizu Y *et al* 2018 *Appl. Phys.* B **124** 71

Chapter 6

Conclusion

This book summarized the role of precision measurements of time and frequency with regard to related developments in physics. In ancient times, the uncertainty in time was of order of 1 h per day, corresponding to a percentage error of 5%–10% (a much larger uncertainty than for length and mass). With this measurement uncertainty, discerning discrepancies when considering Ptolemaic theory was impossible in that era.

The history of the development of clocks and of physics is well correlated. The coincidence cannot go unnoticed between the advances in pendulum clocks and the rise of Newtonian mechanics established in the same period. More recently, the theory of relativity followed on from precise measurements of the speed of light, the universality of which this theory is based on.

Currently, with the development of atomic clocks, the measurement uncertainty in time and frequency is reduced to a level of 10^{-18}, enabling measurement uncertainties for all physical values to be narrowed down to those of time and frequency. As noted in chapter 4, there are still many mysteries that we cannot explain to this day. The role of atomic clocks is expected to be much more important than in the past. With a focus of universal reproducibility in metrological science, changes are to be made in defining the ampere and mole that make up the International System of Units. Moreover, in chapter 5, precise measurements of time and frequency were shown to contribute directly to coordinate many aspects of human life and the interconnectedness of the modern world. The promise of higher accuracy and stability will see a continuance of this trend as machine learning enters the fray of daily life.

As an author, it would be a great pleasure if this book kindles an interest in young researchers for the role that precision measurements of time and frequency have played and will continue to play in the future of mankind.

doi:10.1088/978-0-7503-2124-2ch6

IOP Publishing

Measuring Time
Frequency measurements and related developments in physics
Masatoshi Kajita

Appendix A

Laser cooling

Gaseous atoms move with speeds in the range of 200–400 m s^{-1} (with He, 1000 m s^{-1}). Laser cooling is a technique to decelerate atoms to speeds slower than 10 cm s^{-1} and has had a significant effect on fundamental physics including precision measurements of frequency.

To begin, we consider an atom with one ground state and one excited state. When atoms absorb laser light, the atoms absorb not only the energy but also the momentum of the photon. Therefore, atoms are transformed to an excited energy state and simultaneously experience a force along the direction of light propagation [1]. Afterwards, atoms emit photons in random directions (spontaneous emission) and return to the ground state. With the emission of photons, atoms receive a recoil force; nevertheless, it is randomly directed and hence averages to zero. Repeating the cycle of absorption and spontaneous emission, atoms experience a net force along the direction of light propagation. An atom decelerates when its motion opposes that of light. Note also that light accelerates atoms, when the direction of light propagation is parallel to the atomic motion.

To cool atoms of random velocities, two laser beams of frequency lower than the transition frequency irradiate the atoms from opposing directions. Because of the first-order Doppler effect, atoms interact only with opposing light; all atoms are decelerated (figure A1). As the atoms decelerate and the first-order Doppler shift becomes smaller than the transition linewidth (Γ), the difference in the interaction forces from the two opposing beams becomes smaller. Cooling stops when the kinetic temperature, called the Doppler limit, is reached; the temperature is given by

$$T_{\text{Doppler}} = \frac{h\Gamma}{2k_{\text{B}}}. \tag{A.1}$$

Using a strong transition for laser cooling, the kinetic energy reduces rapidly to the Doppler limit (0.1 – 1 mK). Nevertheless, a weak transition has the advantage in

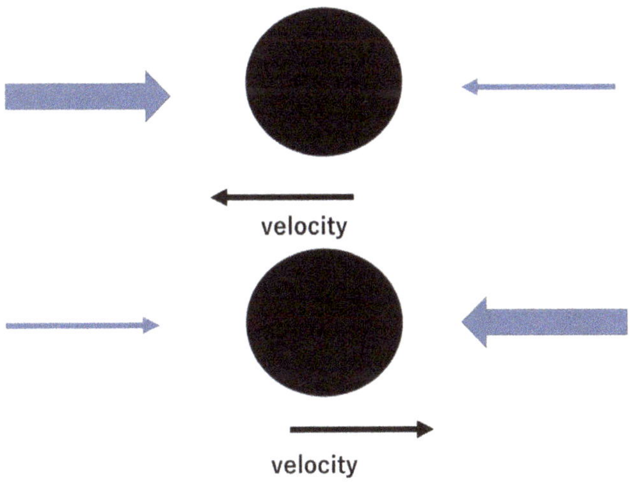

Figure A1. Laser beams, with a frequency below the transition frequency, irradiate the atoms from opposing directions. Atoms interact only with laser light that opposes the motion of the atom.

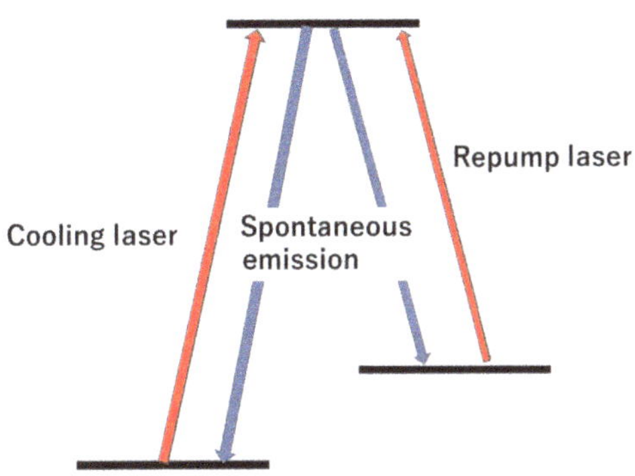

Figure A2. Maintaining the cooling cycle using a repump laser.

being able to attain lower kinetic energies. The best strategy to use is to initially cool the atoms using a strong transition, and subsequently switch to cooling using a weak transition.

We assumed that only a spontaneous emission transition from the excited state to the initial ground state is possible. When a second transition to another state is possible, the cooling cycle (absorption + spontaneous emission) cannot continue for long. A repump laser, giving the transition to the excited state (figure A2) is needed to push atoms back to the cooling cycle. With complex energy states, more repump lasers are required to maintain the laser cooling process. Therefore, laser cooling is possible only for atoms with simple energy structures, such as alkali or alkali-earth atoms. Recently, laser cooling was successfully applied to several molecules with relatively simple energy structures [2, 3]. In a variant of this technique, called sympathetic cooling, the kinetic energy of atoms or molecules can be reduced via collisional interactions with laser-cooled atoms. Sympathetic cooling is useful to

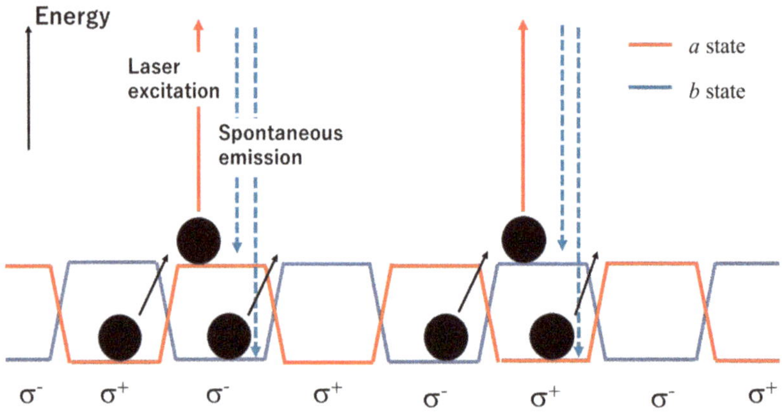

Figure A3. Cooling procedure with polarization-gradient cooling. The degenerate sub-states are split depending on the polarization by the Stark shift induced by the laser light. The laser polarization changes periodically and kinetic energy is lost by the transition between two sub-states.

reduce the kinetic energy of atoms or molecules for which direct laser cooling is difficult.

When laser cooling is performed in one direction, there is a heating effect in the other direction induced by the random force during spontaneous emission. To reduce the kinetic energy in two or three directions, four or six cooling lasers are required. For trapped ions undergoing vibrational motion, laser cooling can be performed with one cooling laser of frequency lower than the transition frequency. Trapped ions repeatedly undergo motion parallel and opposite to the direction of light and interact with the cooling laser only if the motion is in the opposite direction to the beam. The motion of the trapped ions in the three directions are coupled and cooled in all directions by a cooling laser in one direction.

There are several methods to reduce the kinetic energy below the Doppler limit. For example, polarization-gradient cooling is useful for atoms with degenerate ground states: sub-states a and b [4]. This degeneracy is split by the Stark shift induced by the cooling laser and depends on the polarization of the laser (figure A3). With one polarization, the energy of the sub-state a is higher than that of b. Via the absorption transition from a to the excited state and the spontaneous emission transition to b, there is a loss of kinetic energy for the atom. With the other polarization, the energy of a is lower than b and the loss of kinetic energy for the atom arises from the transition from b to a. The two cooling lasers with opposite directions form a standing wave and both polarizations appear periodically. The cooling continues until the kinetic energy of the atoms falls below the energy difference between both sub-states. With this method applied to Cs and Rb atoms, a kinetic temperature of order of a few μK is obtained.

For trapped atoms or ions with vibrational motion of frequency f_{vib}, the energy of motion E_{vib} is given by

$$E_{\text{vib}}(n_v) = \left(n_v + \frac{1}{2}\right)hf_{\text{vib}}. \tag{A.2}$$

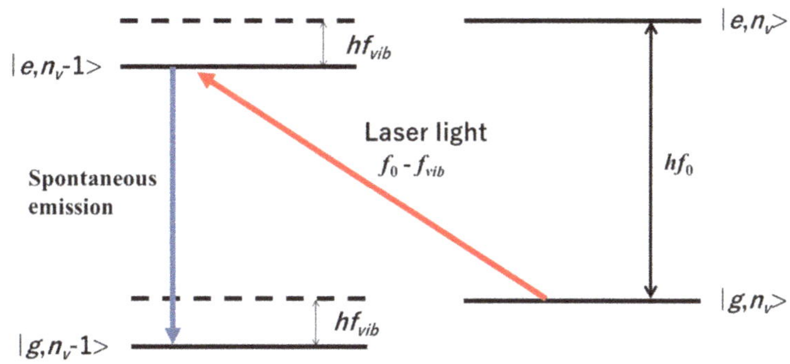

Figure A4. The concept of sideband cooling. Here, f_0 is the transition frequency between the ground (g) and excited state (e); the energy of motion of the trapped atom (vibrational frequency f_{vib}), ion, and molecules is given by $(n_v + 1/2)\ hf_{\text{vib}}$.

A first-order Doppler effect with a narrow linewidth is observed as a sideband of frequency

$$f_s(N_v) = f_0 + N_v f_{\text{vib}} \tag{A.3}$$

where f_0 is the real transition frequency. When an atomic transition is induced by the laser light with $f_s(N_v)$, the energy of motion changes by $N_v hf_{\text{vib}}$. A method called 'sideband cooling' exists that reduces the energy of motion by hf_{vib} per cycle by irradiating the atoms with laser light of frequency $f_s(-1)$; see figure A4 [5]. The transition rates induced by the sideband $f_s(-1)$ and $f_v(1)$ are proportional to n_v and $(n_v + 1)$, respectively. With this method, the energy of motion is reduced to the ground state $hf_{\text{vib}}/2$, called the zero-point energy, from where the transition rate induced by the sideband $f_v(-1)$ is zero. There is another method, called sideband Raman cooling [6], which converts the same ground state to a state of lower energy of motion using two laser beams with a frequency difference of f_{vib}. The zero-point energy can be reduced by decreasing f_{vib} adiabatically, after the energy of motion is reduced to the ground state (adiabatic cooling).

References

[1] Haensch T W and Shawlow A L 1975 *Opt. Commun.* **13** 68
[2] Griffith W C *et al* 2009 *Phys. Rev. Lett.* **102** 101601
[3] Kozyryef I *et al* 2017 *Phys. Rev. Lett.* **119** 133002
[4] Dalibard J and Cohen-Tannoudji C 1989 *J. Opt. Soc. Am.* B **6** 2023
[5] Morigi G *et al* 1999 *Phys. Rev.* A **59** 3797
[6] Hamann S E *et al* 1998 *Phys. Rev. Lett.* **80** 4149

Lightning Source UK Ltd.
Milton Keynes UK
UKHW050646060619
343842UK00004B/103/P

9 780750 321228